GRAVITY AND MAGNETIC FOR GEOLOGICAL STUDIES

GRAVITY AND MAGNETIC FOR GEOLOGICAL STUDIES

Ganesh Prasad

RANDOM PUBLICATIONS

NEW DELHI - 110 002 (INDIA)

Gravity and Magnetic for Geological Studies

ISBN 978-93-51117-07-0

Published in 2015 in India by

RANDOM PUBLICATIONS

4376-A/4B, Gali Murari Lal, Ansari Road
New Delhi-110 002
Phone: +9111-43580356, 23289044
E-mail: randomexports@gmail.com; sales@randompublications.com; info@randompublications.com

Reprint 2021

Type Setting by: Friends Media, Delhi-110089
Printed at : Replika Press Pvt. Ltd.

Preface

Geology gives insight into the history of the Earth by providing the primary evidence for plate tectonics, the evolutionary history of life, and past climates. In modern times, geology is commercially important for mineral and hydrocarbon exploration / exploitation as well as for evaluating water resources. It is publicly important for the prediction and understanding of natural hazards, the remediation of environmental problems, and for providing insights into past climate change. Gravitation or gravity is a natural phenomenon by which all physical bodies attract each other. Gravity gives weight to physical objects and causes them to fall towards the ground when dropped. In modern physics, gravitation is most accurately described by the general theory of relativity (proposed by Einstein) which describes gravitation as a consequence of the curvature of spacetime. For most situations gravity is well approximated by Newton's law of universal gravitation, which postulates that the gravitational force of two bodies of mass is directly proportional to the product of their masses and inversely proportional to the square of the distance between them.

Gravity is the weakest of the four fundamental forces of nature. The gravitational force is approximately 10^{-38} times the strength of the strong force (i.e. gravity is 38 orders of magnitude weaker), $10^{"36}$ times the strength of the electromagnetic force, and 10^{-29} times the strength of the weak force. As a consequence, gravity has a negligible influence on the behaviour of sub-atomic particles, and plays no role in determining the internal properties of everyday matter. On the other hand, gravity is the dominant force at the macroscopic scale, that is the cause of the formation, shape, and trajectory (orbit) of astronomical bodies, including those of asteroids, comets, planets, stars, and galaxies. It is responsible for causing the Earth and the other planets to orbit the Sun; for causing the Moon to orbit the Earth; for the formation of tides; for natural convection, by which fluid flow occurs under the influence of a density gradient and gravity; for heating the interiors of forming stars and planets to very high temperatures; for solar system, galaxy, stellar formation and evolution; and for various other phenomena observed on Earth and throughout the universe. Magnetism is a class of physical phenomena that are mediated by magnetic fields. Electric currents and the fundamental magnetic moments of elementary particles give rise to a magnetic field, which acts on other currents and magnetic moments. All materials are influenced to some extent by a magnetic field. The most familiar effect is on permanent magnets, which have persistent magnetic moments caused by ferromagnetism. Most materials do not have permanent moments. Some are attracted to a magnetic field (paramagnetism); others are repulsed by a magnetic field (diamagnetism); others have a much more complex relationship with an applied magnetic field. Magnetic effects result primarily from the magnetization induced in susceptible rocks by the Earth's magnetic field. Most sedimentary rocks have very low susceptibility and thus are nearly transparent to magnetism. Accordingly, in petroleum exploration magnetics are used negatively: magnetic anomalies indicate the absence of explorable sedimentary rocks. Gravitomagnetism is a widely used term referring specifically to the kinetic effects of gravity, in analogy to the magnetic effects of moving electric charge. General Magnetic and gravity methods have much in common, but magnetics is generally more complex and variations in the magnetic field are more erratic and localized.

This is an ideal text for advanced undergraduate and graduate courses and reference text for research academics and professional geophysicists.

I thank all members of my team who have helped in the preparation of the book. My special thanks go to "Random Publications" who have published the book.

— *Ganesh Prasad*

Contents

Chapter 1

Introduction

Geology is an earth science comprising the study of solid Earth, the rocks of which it is composed, and the processes by which they change.

Geology can also refer generally to the study of the solid features of any celestial body (such as the geology of the Moon or Mars).

Geology gives insight into the history of the Earth by providing the primary evidence for plate tectonics, the evolutionary history of life, and past climates.

In modern times, geology is commercially important for mineral and hydrocarbon exploration / exploitation as well as for evaluating water resources.

It is publicly important for the prediction and understanding of natural hazards, the remediation of environmental problems, and for providing insights into past climate change.

Geology also plays a role in geotechnical engineering and is a major academic discipline.

Geologic Time

The geologic time scale encompasses the history of the Earth. It is bracketed at the old end by the dates of the earliest solar system material at 4.567 Ga, (gigaannum: billion years ago) and the age of the Earth at 4.54 Ga at the beginning of the informally recognised Hadean eon. At the young end of the scale, it is bracketed by the present day in the Holocene epoch.

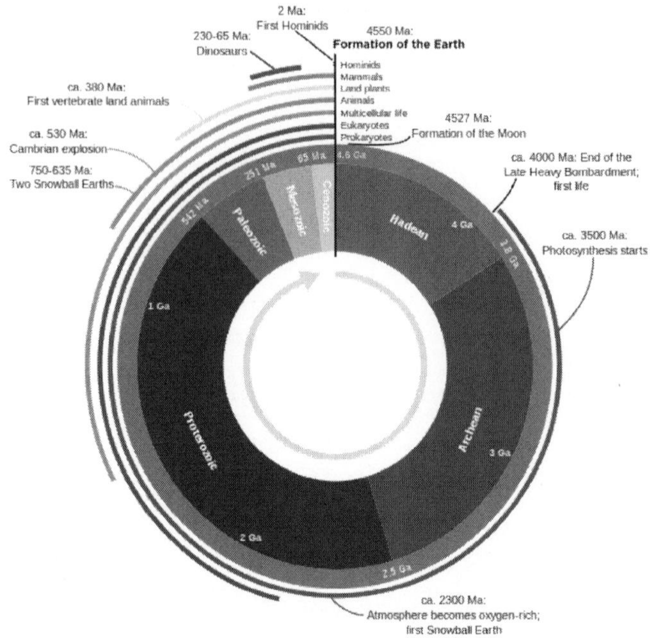

Figure: *Geological time put in a diagram called a geological clock, showing the relative lengths of the eons of the Earth's history.*

Important Milestones

- 4.567 Ga: Solar system formation
- 4.54 Ga: Accretion of Earth
- c. 4 Ga: End of Late Heavy Bombardment, first life
- c. 3.5 Ga: Start of photosynthesis
- c. 2.3 Ga: Oxygenated atmosphere, first snowball Earth
- 730–635 Ma (megaannum: million years ago): second snowball Earth
- 542± 0.3 Ma: Cambrian explosion – vast multiplication of hard-bodied life; first abundant fossils; start of the Paleozoic
- c. 380 Ma: First vertebrate land animals
- 250 Ma: Permian-Triassic extinction – 90% of all land animals die; end of Paleozoic and beginning of Mesozoic
- 66 Ma: Cretaceous–Paleogene extinction – Dinosaurs die; end of Mesozoic and beginning of Cenozoic
- c. 7 Ma: First hominins appear
- 3.9 Ma: First Australopithecus, direct ancestor to modern Homo sapiens, appear

- 200 ka (kiloannum: thousand years ago): First modern Homo sapiens appear in East Africa

Brief Time Scale

The following four timelines show the geologic time scale. The first shows the entire time from the formation of the Earth to the present, but this compresses the most recent eon. Therefore the second scale shows the most recent eon with an expanded scale. The second scale compresses the most recent era, so the most recent era is expanded in the third scale. Since the Quaternary is a very short period with short epochs, it is further expanded in the fourth scale. The second, third, and fourth timelines are therefore each subsections of their preceding timeline as indicated by asterisks. The Holocene (the latest epoch) is too small to be shown clearly on the third timeline on the right, another reason for expanding the fourth scale. The Pleistocene (P) epoch. Q stands for the Quaternary period.

Relative and Absolute Dating

Geological events can be given a precise date at a point in time, or they can be related to other events that came before and after them. Geologists use a variety of methods to give both relative and absolute dates to geological events. They then use these dates to find the rates at which processes occur.

Relative Dating

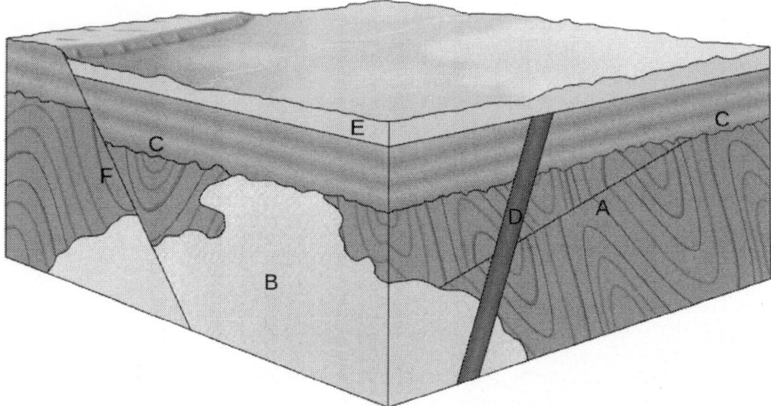

Figure: *Cross-cutting relations can be used to determine the relative ages of rock strata and other geological structures. Explanations: A - folded rock strata cut by a thrust fault; B - large intrusion (cutting through A); C - erosional angular unconformity (cutting off A & B) on which rock strata were deposited; D - volcanic dyke (cutting through A, B & C); E - even younger rock strata (overlying C & D); F - normal fault (cutting through A, B, C & E).*

Methods for relative dating were developed when geology first emerged as a formal science. Geologists still use the following principles today as a means to provide information about geologic history and the timing of geologic events.

The principle of Uniformitarianism states that the geologic processes observed in operation that modify the Earth's crust at present have worked in much the same way over geologic time. A fundamental principle of geology advanced by the 18th century Scottish physician and geologist James Hutton, is that "the present is the key to the past." In Hutton's words: "the past history of our globe must be explained by what can be seen to be happening now."

The principle of intrusive relationships concerns crosscutting intrusions. In geology, when an igneous intrusion cuts across a formation of sedimentary rock, it can be determined that the igneous intrusion is younger than the sedimentary rock. There are a number of different types of intrusions, including stocks, laccoliths, batholiths, sills and dikes.

The principle of cross-cutting relationships pertains to the formation of faults and the age of the sequences through which they cut. Faults are younger than the rocks they cut; accordingly, if a fault is found that penetrates some formations but not those on top of it, then the formations that were cut are older than the fault, and the ones that are not cut must be younger than the fault. Finding the key bed in these situations may help determine whether the fault is a normal fault or a thrust fault.

The principle of inclusions and components states that, with sedimentary rocks, if inclusions (or *clasts*) are found in a formation, then the inclusions must be older than the formation that contains them. For example, in sedimentary rocks, it is common for gravel from an older formation to be ripped up and included in a newer layer. A similar situation with igneous rocks occurs when xenoliths are found. These foreign bodies are picked up as magma or lava flows, and are incorporated, later to cool in the matrix. As a result, xenoliths are older than the rock which contains them.

The principle of original horizontality states that the deposition of sediments occurs as essentially horizontal beds. Observation of modern marine and non-marine sediments in a wide variety of environments supports this generalization (although cross-bedding is inclined, the overall orientation of cross-bedded units is horizontal).

The principle of superposition states that a sedimentary rock layer in a tectonically undisturbed sequence is younger than the one beneath it and older than the one above it. Logically a younger layer cannot slip beneath a layer previously deposited. This principle allows sedimentary layers to be viewed as a form of vertical time line, a partial or complete record of the time elapsed from deposition of the lowest layer to deposition of the highest bed.

The principle of faunal succession is based on the appearance of fossils in sedimentary rocks. As organisms exist at the same time period throughout the world, their presence or (sometimes) absence may be used to provide a relative age of the formations in which they are found. Based on principles laid out by William Smith almost a hundred years before the publication of Charles Darwin's theory of evolution, the principles of succession were developed independently of evolutionary thought.

The principle becomes quite complex, however, given the uncertainties of fossilization, the localization of fossil types due to lateral changes in habitat (facies change in sedimentary strata), and that not all fossils may be found globally at the same time.

Absolute Dating

Geologists can also give precise absolute dates to geologic events. These dates are useful on their own and may also be used in conjunction with relative dating methods or to calibrate relative methods.

At the beginning of the 20th century, the large advance in geology was the ability to give precise absolute dates to geologic events through radioactive isotopes and other methods. The advent of radiometric dating changed the understanding of geologic time. Before, geologists could only use fossils to date sections of rock relative to one another. With isotopic dates, absolute dating became possible, and these absolute dates could be applied to fossil sequences in which there was datable material, converting the old relative ages into new absolute ages.

For many geologic applications, isotope ratios are measured in minerals that give the amount of time that has passed since a rock passed through its particular closure temperature, the point at which different radiometric isotopes stop diffusing into and out of the crystal lattice. These are used in geochronologic and thermochronologic studies. Common methods include uranium-lead dating, potassium-argon dating and argon-argon dating, and uranium-thorium dating. These methods are used for a variety of applications. Dating of lavas and ash layers can help to date stratigraphy and calibrate relative dating techniques. These methods can also be used to determine ages of pluton emplacement.

Thermochemical techniques can be used to determine temperature profiles within the crust, the uplift of mountain ranges, and paleotopography.

Fractionation of the lanthanide series elements is used to compute ages since rocks were removed from the mantle.

Other methods are used for more recent events. Optically stimulated luminescence and cosmogenic radionucleide dating are used to date surfaces and/or erosion rates. Dendrochronology can also be used for the dating of landscapes. Radiocarbon dating is used for young organic material.

Geologic Materials

The majority of geological data comes from research on solid Earth materials. These typically fall into one of two categories: rock and unconsolidated material.

Rock

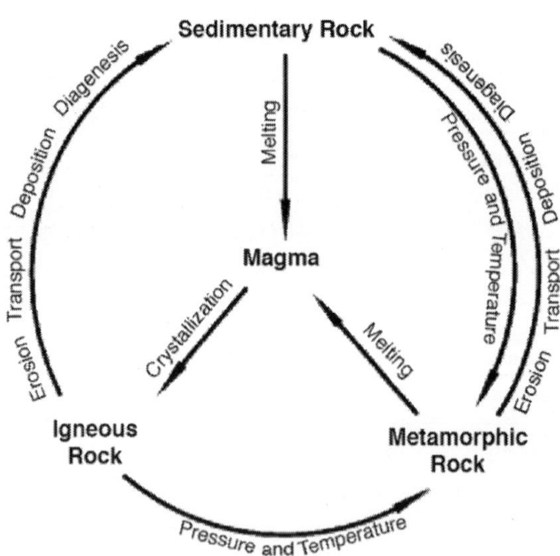

Figure: *This schematic diagram of the rock cycle shows the relationship between magma and sedimentary, metamorphic, and igneous rock*

There are three major types of rock: igneous, sedimentary, and metamorphic. The rock cycle is an important concept in geology which illustrates the relationships between these three types of rock, and magma. When a rock crystallizes from melt (magma and/or lava), it is an igneous rock. This rock can be weathered and eroded, and then redeposited and lithified into a sedimentary rock, or be turned into a metamorphic rock due to heat and pressure that change the mineral content of the rock which gives it a characteristic fabric.

The sedimentary rock can then be subsequently turned into a metamorphic rock due to heat and pressure and is then weathered, eroded, deposited, and lithified, ultimately becoming a sedimentary rock. Sedimentary rock may also be re-eroded and redeposited, and metamorphic rock may also undergo additional metamorphism. All three types of rocks may be re-melted; when this happens, a new magma is formed, from which an igneous rock may once again crystallize.

The majority of research in geology is associated with the study of rock, as rock provides the primary record of the majority of the geologic history of the Earth.

Unconsolidated Material

Geologists also study unlithified material, which typically comes from more recent deposits. Because of this, the study of such material is often known as Quaternary geology, after the recent Quaternary Period. This includes the study of sediment and soils, including studies in geomorphology, sedimentology, and paleoclimatology.

Whole-Earth Structure

Plate Tectonics

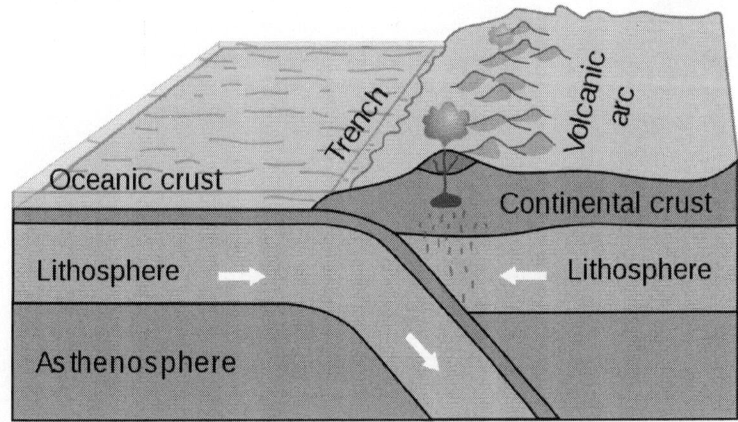

Figure: *Oceanic-continental convergence resulting in subduction and volcanic arcs illustrates one effect of plate tectonics.*

In the 1960s, a series of discoveries, the most important of which was seafloor spreading, showed that the Earth's lithosphere, which includes the crust and rigid uppermost portion of the upper mantle, is separated into a number of tectonic plates that move across the plastically deforming, solid, upper mantle, which is called the asthenosphere. There is an intimate coupling between the movement of the plates on the surface and the convection of the mantle: oceanic plate motions and mantle convection currents always move in the same direction, because the oceanic lithosphere is the rigid upper thermal boundary layer of the convecting mantle. This coupling between rigid plates moving on the surface of the Earth and the convecting mantle is called plate tectonics.

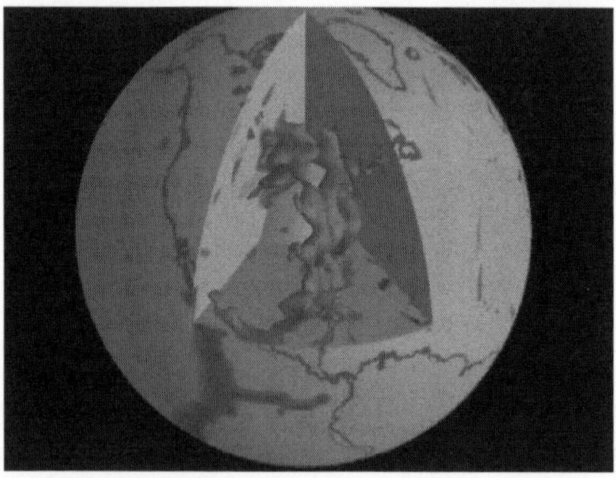

Figure: *On this diagram, subducting slabs are in blue, and continental margins and a few plate boundaries are in red. The blue blob in the cutaway section is the seismically imaged Farallon Plate, which is subducting beneath North America. The remnants of this plate on the Surface of the Earth are the Juan de Fuca Plate and Explorer plate in the Northwestern USA / Southwestern Canada, and the Cocos Plate on the west coast of Mexico.*

The development of plate tectonics provided a physical basis for many observations of the solid Earth. Long linear regions of geologic features could be explained as plate boundaries. Mid-ocean ridges, high regions on the seafloor where hydrothermal vents and volcanoes exist, were explained as divergent boundaries, where two plates move apart. Arcs of volcanoes and earthquakes were explained as convergent boundaries, where one plate subducts under another. Transform boundaries, such as the San Andreas fault system, resulted in widespread powerful earthquakes. Plate tectonics also provided a mechanism for Alfred Wegener's theory of continental drift, in which the continents move across the surface of the Earth over geologic time. They also provided a driving force for crustal deformation, and a new setting for the observations of structural geology. The power of the theory of plate tectonics lies in its ability to combine all of these observations into a single theory of how the lithosphere moves over the convecting mantle.

Earth Structure

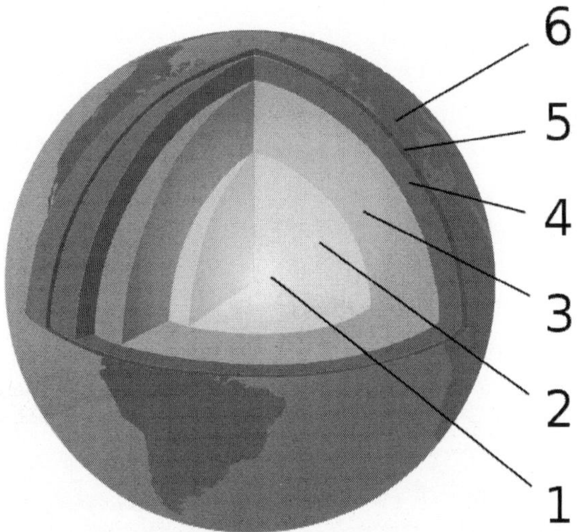

***Figure:** The Earth's layered structure. (1) inner core; (2) outer core; (3) lower mantle; (4) upper mantle; (5) lithosphere; (6) crust (part of the lithosphere)*

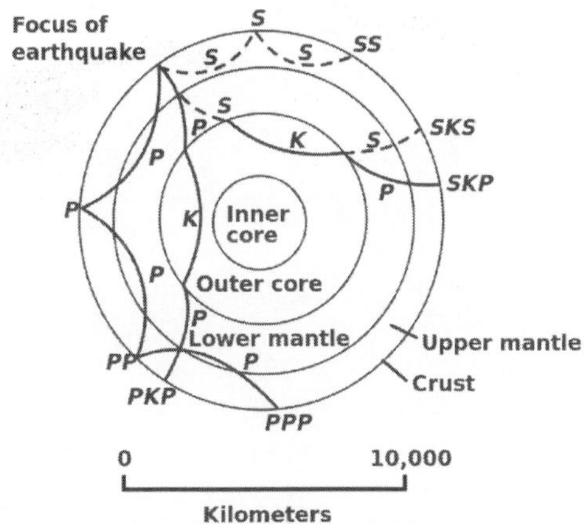

***Figure:** Earth layered structure. Typical wave paths from earthquakes like these gave early seismologists insights into the layered structure of the Earth*

Advances in seismology, computer modelling, and mineralogy and crystallography at high temperatures and pressures give insights into the internal composition and structure of the Earth.

Seismologists can use the arrival times of seismic waves in reverse to image the interior of the Earth. Early advances in this field showed the existence of a liquid outer core (where shear waves were not able to propagate) and a dense solid inner core.

These advances led to the development of a layered model of the Earth, with a crust and lithosphere on top, the mantle below (separated within itself by seismic discontinuities at 410 and 660 kilometres), and the outer core and inner core below that.

More recently, seismologists have been able to create detailed images of wave speeds inside the earth in the same way a doctor images a body in a CT scan. These images have led to a much more detailed view of the interior of the Earth, and have replaced the simplified layered model with a much more dynamic model.

Mineralogists have been able to use the pressure and temperature data from the seismic and modelling studies alongside knowledge of the elemental composition of the Earth to reproduce these conditions in experimental settings and measure changes in crystal structure. These studies explain the chemical changes associated with the major seismic discontinuities in the mantle and show the crystallographic structures expected in the inner core of the Earth.

Geological Development of An Area

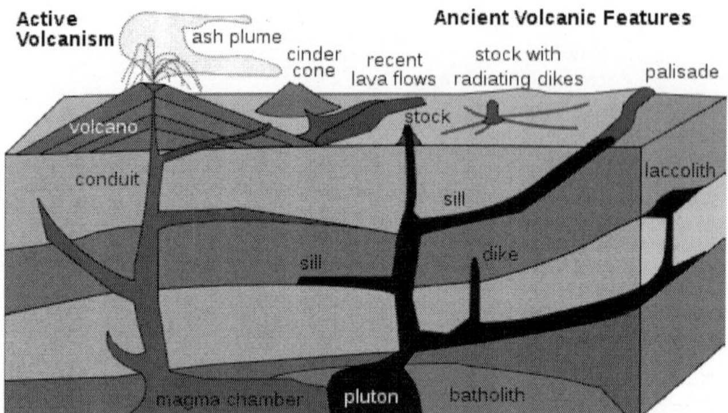

Figure: *An originally horizontal sequence of sedimentary rocks (in shades of tan) are affected by igneous activity. Deep below the surface are a magma chamber and large associated igneous bodies. The magma chamber feeds the volcano, and sends off shoots of magma that will later crystallize into dikes and sills. Magma also advances upwards to form intrusive igneous bodies. The diagram illustrates both a cinder cone volcano, which releases ash, and a composite volcano, which releases both lava and ash.*

The geology of an area changes through time as rock units are deposited and inserted and deformational processes change their shapes and locations.

Figure: *An illustration of the three types of faults. Strike-slip faults occur when rock units slide past one another, normal faults occur when rocks are undergoing horizontal extension, and thrust faults occur when rocks are undergoing horizontal shortening.*

Rock units are first emplaced either by deposition onto the surface or intrusion into the overlying rock. Deposition can occur when sediments settle onto the surface of the Earth and later lithify into sedimentary rock, or when as volcanic material such as volcanic ash or lava flows blanket the surface. Igneous intrusions such as batholiths, laccoliths, dikes, and sills, push upwards into the overlying rock, and crystallize as they intrude.

After the initial sequence of rocks has been deposited, the rock units can be deformed and/or metamorphosed. Deformation typically occurs as a result of horizontal shortening, horizontal extension, or side-to-side (strike-slip) motion. These structural regimes broadly relate to convergent boundaries, divergent boundaries, and transform boundaries, respectively, between tectonic plates.

When rock units are placed under horizontal compression, they shorten and become thicker. Because rock units, other than muds, do not significantly change in volume, this is accomplished in two primary ways: through faulting and folding. In the shallow crust, where brittle deformation can occur, thrust faults form, which cause deeper rock to move on top of shallower rock. Because deeper rock is often older, as noted by the principle of superposition, this can result in older rocks moving on top of younger ones. Movement along faults can result in folding, either because the faults are not planar or because rock layers are dragged along, forming drag folds as slip occurs along the fault. Deeper in the Earth, rocks behave plastically, and fold instead of faulting. These folds can either be those where the material in the centre of the fold buckles upwards, creating "antiforms", or where it buckles downwards, creating "synforms". If the tops of the rock units within the folds remain pointing upwards, they are called anticlines and synclines, respectively. If some of the units in the fold are facing downward, the structure is called an overturned anticline or syncline, and if all of the rock units are overturned or the correct up-direction is unknown, they are simply called by the most general terms, antiforms and synforms.

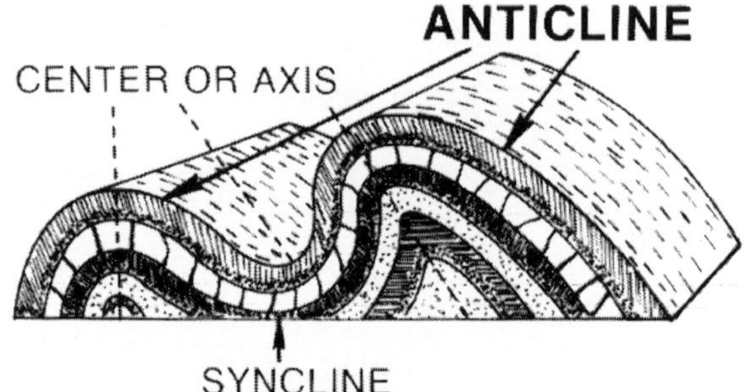

Figure: *A diagram of folds, indicating an anticline and a syncline.*

Even higher pressures and temperatures during horizontal shortening can cause both folding and metamorphism of the rocks. This metamorphism causes changes in the mineral composition of the rocks; creates a foliation, or planar surface, that is related to mineral growth under stress. This can remove signs of the original textures of the rocks, such as bedding in sedimentary rocks, flow features of lavas, and crystal patterns in crystalline rocks.

Extension causes the rock units as a whole to become longer and thinner. This is primarily accomplished through normal faulting and through the ductile stretching and thinning. Normal faults drop rock units that are higher below those that are lower. This typically results in younger units being placed below older units. Stretching of units can result in their thinning; in fact, there is a location within the Maria Fold and Thrust Belt in which the entire sedimentary sequence of the Grand Canyon can be seen over a length of less than a metre. Rocks at the depth to be ductilely stretched are often also metamorphosed. These stretched rocks can also pinch into lenses, known as boudins, after the French word for "sausage", because of their visual similarity.

Where rock units slide past one another, strike-slip faults develop in shallow regions, and become shear zones at deeper depths where the rocks deform ductilely.

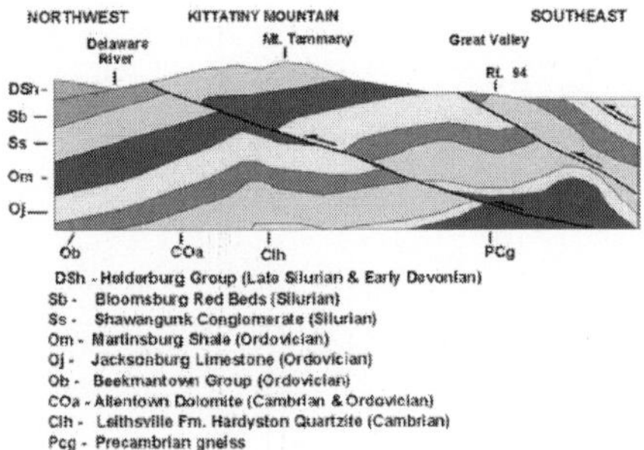

Figure: *Geologic cross-section of Kittatinny Mountain. This cross-section shows metamorphic rocks, overlain by younger sediments deposited after the metamorphic event. These rock units were later folded and faulted during the uplift of the mountain.*

The addition of new rock units, both depositionally and intrusively, often occurs during deformation. Faulting and other deformational processes result in the creation of topographic gradients, causing material on the rock unit that is increasing in elevation to be eroded by hillslopes and channels. These sediments are deposited on the rock unit that is going down. Continual motion along the fault maintains the topographic gradient in spite of the movement of sediment, and continues to create accommodation space for the material to deposit. Deformational events are often also associated with volcanism and igneous activity. Volcanic ashes and lavas accumulate on the surface, and igneous intrusions enter from below. Dikes, long, planar igneous intrusions, enter along cracks, and therefore often form in large numbers in areas that are being actively deformed. This can result in the emplacement of dike swarms, such as those that are observable across the Canadian shield, or rings of dikes around the lava tube of a volcano.

All of these processes do not necessarily occur in a single environment, and do not necessarily occur in a single order. The Hawaiian Islands, for example, consist almost entirely of layered basaltic lava flows. The sedimentary sequences of the mid-continental United States and the Grand Canyon in the southwestern United States contain almost-undeformed stacks of sedimentary rocks that have remained in place since Cambrian time. Other areas are much more geologically complex. In the southwestern United States, sedimentary, volcanic, and intrusive rocks have been metamorphosed, faulted, foliated, and folded. Even older rocks, such as the Acasta gneiss of the

Slave craton in northwestern Canada, the oldest known rock in the world have been metamorphosed to the point where their origin is undiscernable without laboratory analysis. In addition, these processes can occur in stages. In many places, the Grand Canyon in the southwestern United States being a very visible example, the lower rock units were metamorphosed and deformed, and then deformation ended and the upper, undeformed units were deposited. Although any amount of rock emplacement and rock deformation can occur, and they can occur any number of times, these concepts provide a guide to understanding the geological history of an area.

Gravitation

Gravitation or gravity is a natural phenomenon by which all physical bodies attract each other. Gravity gives weight to physical objects and causes them to fall toward the ground when dropped.

In modern physics, gravitation is most accurately described by the general theory of relativity (proposed by Einstein) which describes gravitation as a consequence of the curvature of spacetime. For most situations gravity is well approximated by Newton's law of universal gravitation, which postulates that the gravitational force of two bodies of mass is directly proportional to the product of their masses and inversely proportional to the square of the distance between them.

In pursuit of a theory of everything, the merging of general relativity and quantum mechanics (or quantum field theory) into a more general theory of quantum gravity has become an area of active research. It is hypothesised that the gravitational force is mediated by a massless spin-2 particle called the graviton, and that gravity would have separated from the electronuclear force during the grand unification epoch.

Gravity is the weakest of the four fundamental forces of nature. The gravitational force is approximately 10^{-38} times the strength of the strong force (i.e. gravity is 38 orders of magnitude weaker), 10^{-36} times the strength of the electromagnetic force, and 10^{-29} times the strength of the weak force. As a consequence, gravity has a negligible influence on the behaviour of sub-atomic particles, and plays no role in determining the internal properties of everyday matter. On the other hand, gravity is the dominant force at the macroscopic scale, that is the cause of the formation, shape, and trajectory (orbit) of astronomical bodies, including those of asteroids, comets, planets, stars, and galaxies. It is responsible for causing the Earth and the other planets to orbit the Sun; for causing the Moon to orbit the Earth; for the formation of tides; for natural convection, by which fluid flow occurs under the influence of a density gradient and gravity; for heating the interiors of forming stars and planets to very high temperatures; for solar system, galaxy, stellar formation and evolution; and for various other phenomena observed on Earth and throughout the universe. This is the case for several reasons: gravity is the only force acting on all particles with mass; it has an infinite range; it is always attractive and never repulsive; and it cannot be absorbed, transformed, or shielded against. Even though electromagnetism is far stronger than gravity, electromagnetism is not relevant to astronomical objects, since such bodies have an equal number of protons and electrons that cancel out (i.e., a net electric charge of zero).

Around fourth century CE, the famous Hindu astronomer, Bhâskara II in his work Surya Siddhanta wrote: "Objects fall on the earth due to a force of attraction by the earth. Therefore, the earth, planets, constellations, moon and sun are held in orbit due to this attraction." Newton rediscovered this 1200 years later.

Scientific Revolution

Modern work on gravitational theory began with the work of Galileo Galilei in the late 16th and early 17th centuries. In his famous (though possibly apocryphal) experiment dropping balls from the Tower of Pisa, and later with careful measurements of balls rolling down inclines, Galileo showed that gravitation accelerates all objects at the same rate. This was a major departure from Aristotle's belief that heavier objects accelerate faster. Galileo postulated air resistance as the reason that lighter objects may fall slower in an atmosphere. Galileo's work set the stage for the formulation of Newton's theory of gravity.

Newton's Theory of Gravitation

In 1687, English mathematician Sir Isaac Newton published *Principia*, which hypothesizes the inverse-square law of universal gravitation. In his own words, "I deduced that the forces which keep the planets in their orbs must [be] reciprocally as the squares of their distances from the centres about which they revolve: and thereby compared the force requisite to keep the Moon in her Orb with the force of gravity at the surface of the Earth; and found them answer pretty nearly."

Newton's theory enjoyed its greatest success when it was used to predict the existence of Neptune based on motions of Uranus that could not be accounted for by the actions of the other planets. Calculations by both John Couch Adams and Urbain Le Verrier predicted the general position of the planet, and Le Verrier's calculations are what led Johann Gottfried Galle to the discovery of Neptune.

A discrepancy in Mercury's orbit pointed out flaws in Newton's theory. By the end of the 19th century, it was known that its orbit showed slight perturbations that could not be accounted for entirely under Newton's theory, but all searches for another perturbing body (such as a planet orbiting the Sun even closer than Mercury) had been fruitless. The issue was resolved in 1915 by Albert Einstein's new theory of general relativity, which accounted for the small discrepancy in Mercury's orbit.

Although Newton's theory has been superseded, most modern non-relativistic gravitational calculations are still made using Newton's theory because it is a much simpler theory to work with than general relativity, and gives sufficiently accurate results for most applications involving sufficiently small masses, speeds and energies.

Equivalence Principle

The equivalence principle, explored by a succession of researchers including Galileo, Loránd Eötvös, and Einstein, expresses the idea that all objects fall in the same way. The simplest way to test the weak equivalence principle is to drop two objects of different masses or compositions in a vacuum and see whether they hit the ground at the same time. Such experiments demonstrate that all objects fall at the same rate when friction (including air resistance) is negligible. More sophisticated tests use a torsion balance of a type invented by Eötvös. Satellite experiments, for example STEP, are planned for more accurate experiments in space.

Formulations of the equivalence principle include:

- The weak equivalence principle: *The trajectory of a point mass in a gravitational field depends only on its initial position and velocity, and is independent of its composition.*
- The Einsteinian equivalence principle: *The outcome of any local non-gravitational experiment in a freely falling laboratory is independent of the velocity of the laboratory and its location in spacetime.*
- The strong equivalence principle requiring both of the above.

General Relativity

In general relativity, the effects of gravitation are ascribed to spacetime curvature instead of a force. The starting point for general relativity is the equivalence principle, which equates free fall with inertial motion and describes free-falling inertial objects as being accelerated relative to non-inertial observers on the ground. In Newtonian physics, however, no such acceleration can occur unless at least one of the objects is being operated on by a force.

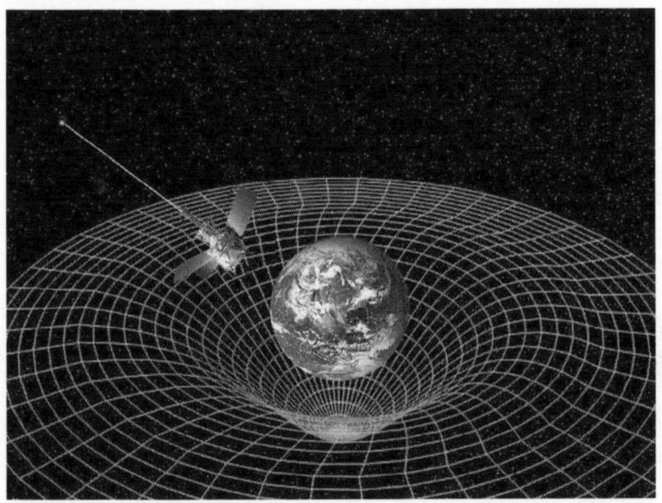

Figure: *Two-dimensional analogy of spacetime distortion generated by the mass of an object. Matter changes the geometry of spacetime, this (curved) geometry being interpreted as gravity. White lines do not represent the curvature of space but instead represent the coordinate system imposed on the curved spacetime, which would be rectilinear in a flat spacetime.*

Einstein proposed that spacetime is curved by matter, and that free-falling objects are moving along locally straight paths in curved spacetime. These straight paths are called geodesics. Like Newton's first law of motion, Einstein's theory states that if a force is applied on an object, it would deviate from a geodesic.

For instance, we are no longer following geodesics while standing because the mechanical resistance of the Earth exerts an upward force on us, and we are non-inertial on the ground as a result. This explains why moving along the geodesics in spacetime is considered inertial.

Einstein discovered the field equations of general relativity, which relate the presence of matter and the curvature of spacetime and are named after him.

The Einstein field equations are a set of 10 simultaneous, non-linear, differential equations. The solutions of the field equations are the components of the metric tensor of spacetime. A metric tensor describes a geometry of spacetime. The geodesic paths for a spacetime are calculated from the metric tensor.

Notable Solutions of the Einstein Field Equations Include

- The Schwarzschild solution, which describes spacetime surrounding a spherically symmetric non-rotating uncharged massive object.

  For compact enough objects, this solution generated a black hole with a central singularity.

  For radial distances from the centre which are much greater than the Schwarzschild radius, the accelerations predicted by the Schwarzschild solution are practically identical to those predicted by Newton's theory of gravity.
- The Reissner-Nordström solution, in which the central object has an electrical charge. For charges with a geometrized length which are less than the geometrized length of the mass of the object, this solution produces black holes with two event horizons.
- The Kerr solution for rotating massive objects. This solution also produces black holes with multiple event horizons.
- The Kerr-Newman solution for charged, rotating massive objects. This solution also produces black holes with multiple event horizons.
- The cosmological Friedmann-Lemaître-Robertson-Walker solution, which predicts the expansion of the universe.

The Tests of General Relativity Included the Following

- General relativity accounts for the anomalous perihelion precession of Mercury.
- The prediction that time runs slower at lower potentials has been confirmed by the Pound–Rebka experiment, the Hafele–Keating experiment, and the GPS.
- The prediction of the deflection of light was first confirmed by Arthur Stanley Eddington from his observations during the Solar eclipse of May 29, 1919. Eddington measured starlight deflections twice those predicted by Newtonian corpuscular theory, in accordance with the predictions of general relativity.

  However, his interpretation of the results was later disputed. More recent tests using radio interferometric measurements of quasars passing behind the Sun have more accurately and consistently confirmed the deflection of light to the degree predicted by general relativity.

- The time delay of light passing close to a massive object was first identified by Irwin I. Shapiro in 1964 in interplanetary spacecraft signals.
- Gravitational radiation has been indirectly confirmed through studies of binary pulsars.
- Alexander Friedmann in 1922 found that Einstein equations have non-stationary solutions (even in the presence of the cosmological constant). In 1927 Georges Lemaître showed that static solutions of the Einstein equations, which are possible in the presence of the cosmological constant, are unstable, and therefore the static universe envisioned by Einstein could not exist. Later, in 1931, Einstein himself agreed with the results of Friedmann and Lemaître. Thus general relativity predicted that the Universe had to be non-static—it had to either expand or contract. The expansion of the universe discovered by Edwin Hubble in 1929 confirmed this prediction.
- The theory's prediction of frame dragging was consistent with the recent Gravity Probe B results.
- General relativity predicts that light should lose its energy when travelling away from the massive bodies. The group of Radek Wojtak of the Niels Bohr Institute at the University of Copenhagen collected data from 8000 galaxy clusters and found that the light coming from the cluster centres tended to be red-shifted compared to the cluster edges, confirming the energy loss due to gravity.

Gravity and Quantum Mechanics

In the decades after the discovery of general relativity it was realised that general relativity is incompatible with quantum mechanics. It is possible to describe gravity in the framework of quantum field theory like the other fundamental forces, such that the attractive force of gravity arises due to exchange of virtual gravitons, in the same way as the electromagnetic force arises from exchange of virtual photons. This reproduces general relativity in the classical limit. However, this approach fails at short distances of the order of the Planck length, where a more complete theory of quantum gravity (or a new approach to quantum mechanics) is required.

Earth's Gravity

Every planetary body (including the Earth) is surrounded by its own gravitational field, which exerts an attractive force on all objects.

Assuming a spherically symmetrical planet, the strength of this field at any given point is proportional to the planetary body's mass and inversely proportional to the square of the distance from the centre of the body.

The strength of the gravitational field is numerically equal to the acceleration of objects under its influence, and its value at the Earth's surface, denoted *g*, is expressed below as the standard average. According to the International Bureau of Weights and Measures, under the International System of Units (SI), the Earth's standard acceleration due to gravity is:

$$g = 9.80665 \text{ m/s}^2 \text{ (32.1740 ft/s}^2\text{)}.$$

This means that, ignoring air resistance, an object falling freely near the Earth's surface increases its velocity by 9.80665 m/s (32.1740 ft/s or 22 mph) for each second of its descent. Thus, an object starting from rest will attain a velocity of 9.80665 m/s (32.1740 ft/s) after one second, approximately 19.62 m/s (64.4 ft/s) after two seconds, and so on, adding 9.80665 m/s (32.1740 ft/s) to each resulting velocity. Also, again ignoring air resistance, any and all objects, when dropped from the same height, will hit the ground at the same time.

According to Newton's 3rd Law, the Earth itself experiences a force equal in magnitude and opposite in direction to that which it exerts on a falling object. This means that the Earth also accelerates towards the object until they collide. Because the mass of the Earth is huge, however, the acceleration imparted to the Earth by this opposite force is negligible in comparison to the object's. If the object doesn't bounce after it has collided with the Earth, each of them then exerts a repulsive contact force on the other which effectively balances the attractive force of gravity and prevents further acceleration.

The force of gravity on Earth is the resultant (vector sum) of two forces: (a) The gravitational attraction in accordance with Newton's universal law of gravitation, and (b) the centrifugal force, which results from the choice of an earthbound, rotating frame of reference. At the equator, the force of gravity is the weakest due to the centrifugal force caused by the Earth's rotation. The force of gravity varies with latitude and becomes stronger as you increase in latitude toward the poles. The standard value of 9.80665 m/s^2 is the one originally adopted by the International Committee on Weights and Measures in 1901 for 45° latitude, even though it has been shown to be too high by about five parts in ten thousand. This value has persisted in meteorology and in some standard atmospheres as the value for 45° latitude even though it applies more precisely to latitude of 45°32'33".

Equations for a Falling Body Near the Surface of the Earth

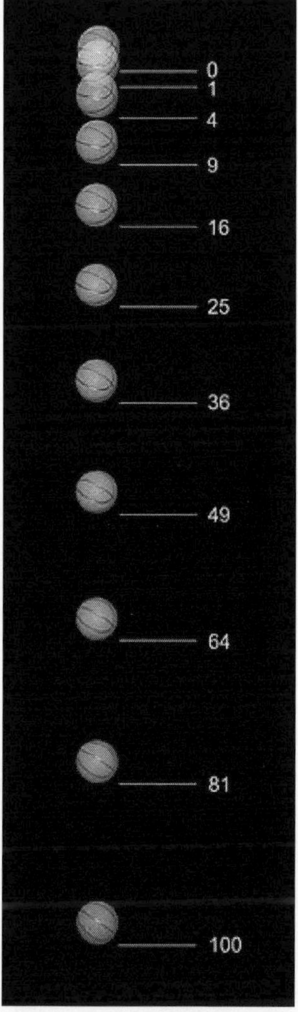

Figure: *Ball falling freely under gravity. See text for description.*

Under an assumption of constant gravity, Newton's law of universal gravitation simplifies to $F = mg$, where m is the mass of the body and g is a constant vector with an average magnitude of 9.81 m/s^2. The acceleration due to gravity is equal to this g. An initially stationary object which is allowed to fall freely under gravity drops a distance which is proportional to the square of the elapsed time. The image on the right, spanning half a second, was captured with a stroboscopic flash at 20 flashes per second. During the first $^1/_{20}$ of a second the ball drops one unit of distance (here, a unit is about 12 mm); by $^2/_{20}$ it has dropped at total of 4 units; by $^3/_{20}$, 9 units and so on.

Under the same constant gravity assumptions, the potential energy, E_p, of a body at height h is given by $E_p = mgh$ (or $E_p = Wh$, with W meaning weight). This expression is valid only over small distances h from the surface of the Earth. Similarly, the expression $h = \frac{v^2}{2g}$ for the maximum height reached by a vertically projected body with initial velocity v is useful for small heights and small initial velocities only.

Gravity and Astronomy

The discovery and application of Newton's law of gravity accounts for the detailed information we have about the planets in our solar system, the mass of the Sun, the distance to stars, quasars and even the theory of dark matter. Although we have not travelled to all the planets nor to the Sun, we know their masses.

These masses are obtained by applying the laws of gravity to the measured characteristics of the orbit. In space an object maintains its orbit because of the force of gravity acting upon it. Planets orbit stars, stars orbit Galactic Centres, galaxies orbit a centre of mass in clusters, and clusters orbit in superclusters. The force of gravity exerted on one object by another is directly proportional to the product of those objects' masses and inversely proportional to the square of the distance between them.

Gravitational Radiation

In general relativity, gravitational radiation is generated in situations where the curvature of spacetime is oscillating, such as is the case with co-orbiting objects. The gravitational radiation emitted by the Solar System is far too small to measure. However, gravitational radiation has been indirectly observed as an energy loss over time in binary pulsar systems such as PSR B1913+16. It is believed that neutron star mergers and black hole formation may create detectable amounts of gravitational radiation. Gravitational radiation observatories such as the Laser Interferometre Gravitational Wave Observatory (LIGO) have been created to study the problem. No confirmed detections have been made of this hypothetical radiation.

Speed of Gravity

In December 2012, a research team in China announced that it had produced measurements of the phase lag of Earth tides during full and new moons which seem to prove that the speed of gravity is equal to the speed of light. The team's findings were released in the Chinese Science Bulletin in February 2013.

Anomalies and Discrepancies

There are some observations that are not adequately accounted for, which may point to the need for better theories of gravity or perhaps be explained in other ways.

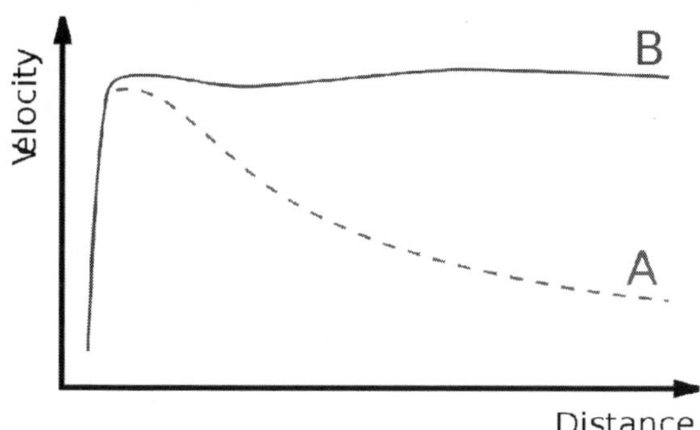

Figure: *Rotation curve of a typical spiral galaxy: predicted (A) and observed (B). The discrepancy between the curves is attributed to dark matter.*

- Extra-fast stars: Stars in galaxies follow a distribution of velocities where stars on the outskirts are moving faster than they should according to the observed distributions of normal matter. Galaxies within galaxy clusters show a similar pattern. Dark matter, which would interact gravitationally but not electromagnetically, would account for the discrepancy. Various modifications to Newtonian dynamics have also been proposed.
- Flyby anomaly: Various spacecraft have experienced greater acceleration than expected during gravity assist maneuvers.
- Accelerating expansion: The metric expansion of space seems to be speeding up. Dark energy has been proposed to explain this. A recent alternative explanation is that the geometry of space is not homogeneous (due to clusters of galaxies) and that when the data are reinterpreted to take this into account, the expansion is not speeding up after all, however this conclusion is disputed.
- Anomalous increase of the astronomical unit: Recent measurements indicate that planetary orbits are widening faster than if this were solely through the sun losing mass by radiating energy.
- Extra energetic photons: Photons travelling through galaxy clusters should gain energy and then lose it again on the way out. The accelerating expansion of the universe should stop the photons returning all the energy, but even taking this into account photons from the cosmic microwave background

radiation gain twice as much energy as expected. This may indicate that gravity falls off *faster* than inverse-squared at certain distance scales.

- Extra massive hydrogen clouds: The spectral lines of the Lyman-alpha forest suggest that hydrogen clouds are more clumped together at certain scales than expected and, like dark flow, may indicate that gravity falls off *slower* than inverse-squared at certain distance scales.

Gauge Gravitation Theory

In quantum field theory, gauge gravitation theory is the effort to extend Yang–Mills theory, which provides a universal description of the fundamental interactions, to describe gravity. It should not be confused with the related but distinct gauge theory gravity.

The first gauge model of gravity was suggested by R. Utiyama in 1956 just two years after birth of the gauge theory itself. However, the initial attempts to construct the gauge theory of gravity by analogy with the gauge models of internal symmetries encountered a problem of treating general covariant transformations and establishing the gauge status of a pseudo-Riemannian metric (a tetrad field).

In order to overcome this drawback, representing tetrad fields as gauge fields of the translation group was attempted. Infinitesimal generators of general covariant transformations were considered as those of the translation gauge group, and a tetrad (coframe) field was identified with the translation part of an affine connection on a world manifold *X*. Any such connection is a sum $K = \Gamma + \Theta$ of a linear world connection Γ and a soldering form $\Theta = \Theta^a_\mu dx^\mu \otimes \vartheta_a$ where $\vartheta_a = \vartheta^\lambda_a \partial_\lambda$ is a non-holonomic frame. For instance, if K is the Cartan connection, then $\Theta = \theta = dx^\mu \otimes \partial_\mu$ is the canonical soldering form on X. There are different physical interpretations of the translation part Θ of affine connections. In gauge theory of dislocations, a field Θ describes a distortion. At the same time, given a linear frame ϑ_a, the decomposition $\theta = \vartheta^a \otimes \vartheta_a$ motivates many authors to treat a coframe ϑ^a as a translation gauge field.

Difficulties of constructing gauge gravitation theory by analogy with the Yang-Mills one result from the gauge transformations in these theories belonging to different classes. In the case of internal symmetries, the gauge transformations are just vertical automorphisms of a principal bundle $P \to X$ leaving its base X fixed. On the other hand, gravitation theory is built on the principal bundle FX of the tangent frames to X. It belongs to the category of natural bundles $T \to X$ for which diffeomorphisms of the base X canonically give rise to automorphisms of T. These automorphisms are called general covariant transformations. General covariant transformations are sufficient in order to restate Einstein's General Relativity and metric-affine gravitation theory as the gauge ones.

In terms of gauge theory on natural bundles, gauge fields are linear connections on a world manifold X, defined as principal connections on the linear frame bundle FX, and a metric (tetrad) gravitational field plays the role of a Higgs field responsible for spontaneous symmetry breaking of general covariant transformations.

Spontaneous symmetry breaking is a quantum effect when the vacuum is not invariant under the transformation group. In classical gauge theory, spontaneous symmetry breaking occurs if the structure group G of a principal bundle $P \to X$ is reducible to a closed subgroup H, i.e., there exists a principal subbundle of P with the structure group H. By virtue of the well-known theorem, there exists one-to-one correspondence between the reduced principal subbundles of P with the structure group H and the global sections of the quotient bundle $P/H \to X$. These sections are treated as classical Higgs fields.

The idea of the pseudo-Riemannian metric as a Higgs field appeared while constructing non-linear (induced) representations of the general linear group $GL(4,\mathbb{R})$, of which the Lorentz group is a Cartan subgroup. The geometric equivalence principle postulating the existence of a reference frame in which Lorentz invariants are defined on the whole world manifold is the theoretical justification of that the structure group $GL(4,\mathbb{R})$ of the linear frame bundle FX is reduced to the Lorentz group. Then the very definition of a pseudo-Riemannian metric on a manifold X as a global section of the quotient bundle $FX/O(1,3) \to X$ leads to its physical interpretation as a Higgs field. The physical reason for world symmetry breaking is the existence of Dirac fermion matter, whose symmetry group is the universal two-sheeted covering $SL(2,\mathbb{C})$ of the restricted Lorentz group.

Chapter 2

Gravitational Wave

Gravitational waves are ripples in the curvature of spacetime that propagate as a wave, travelling outward from the source. Predicted in 1916 by Albert Einstein to exist on the basis of his theory of general relativity, gravitational waves theoretically transport energy as gravitational radiation. Sources of detectable gravitational waves could possibly include binary star systems composed of white dwarfs, neutron stars, or black holes. The existence of gravitational waves is a possible consequence of the Lorentz invariance of general relativity since it brings the concept of a limiting speed of propagation of the physical interactions with it. Gravitational waves cannot exist in the Newtonian theory of gravitation, in which physical interactions propagate at infinite speed.

Although gravitational radiation has not been *directly* detected, there is *indirect* evidence for its existence. For example, the 1993 Nobel Prize in Physics was awarded for measurements of the Hulse–Taylor binary system that suggests gravitational waves are more than mathematical anomalies. Various gravitational wave detectors exist and on 17 March 2014, astronomers at the Harvard–Smithsonian Centre for Astrophysics claimed that they had detected and produced "the first direct image of gravitational waves across the primordial sky" within the cosmic microwave background, providing strong evidence for inflation and the Big Bang.

In Einstein's theory of general relativity, gravity is treated as a phenomenon resulting from the curvature of spacetime. This curvature is caused by the presence of mass. Generally, the more mass that is contained within a given volume of space, the greater the curvature of spacetime will be at the boundary of this volume. As objects with mass move around in spacetime, the curvature changes to reflect the changed locations of those objects. In certain circumstances, accelerating objects generate changes in this curvature, which propagate outwards at the speed of light in a wave-like manner. These propagating phenomena are known as gravitational waves.

As a gravitational wave passes a distant observer, that observer will find spacetime distorted by the effects of strain. Distances between free objects increase and decrease rhythmically as the wave passes, at a frequency corresponding to that of the wave. This occurs despite such free objects never being subjected to an unbalanced force. The magnitude of this effect decreases inversely with distance from the source. Inspiralling binary neutron stars are predicted to be a powerful source of gravitational waves as they coalesce, due to the very large acceleration of their masses as they orbit close to one another. However, due to the astronomical distances to these sources the effects when measured on Earth are predicted to be very small, having strains of less than 1 part in 10^{20}. Scientists are attempting to demonstrate the existence of these waves with ever more sensitive detectors. The current most sensitive measurement is about one part in 5×10^{22} (as of 2012) provided by the LIGO and VIRGO observatories. The lack of detection in these observatories provides an upper limit on the frequency of such powerful sources. A space based observatory, the Laser Interferometre Space Antenna, is currently under development by ESA.

Gravitational waves should penetrate regions of space that electromagnetic waves cannot. It is hypothesized that they will be able to provide observers on Earth with information about black holes and other exotic objects in the distant Universe. Such systems cannot be observed with more traditional means such as optical telescopes and radio telescopes. In particular, gravitational waves could be of interest to cosmologists as they offer a possible way of observing the very early universe. This is not possible with conventional astronomy, since before recombination the universe was opaque to electromagnetic radiation. Precise measurements of gravitational waves will also allow scientists to test the general theory of relativity more thoroughly.

In principle, gravitational waves could exist at any frequency. However, very low frequency waves would be impossible to detect and there is no credible source for detectable waves of very high frequency. Stephen W. Hawking and Werner Israel list different frequency bands for gravitational waves that could be plausibly detected, ranging from 10^{-7} Hz up to 10^{11} Hz.

Effects of a Passing Gravitational Wave

The effects of a passing gravitational wave can be visualized by imagining a perfectly flat region of spacetime with a group of motionless test particles lying in a plane (the surface of your screen). As a gravitational wave passes through the particles along a line perpendicular to the plane of the particles (i.e. following your line of vision into the screen), the particles will follow the distortion in spacetime, oscillating in a "cruciform" manner, as shown in the animations. The area enclosed by the test particles does not change and there is no motion along the direction of propagation.

The oscillations depicted here in the animation are exaggerated for the purpose of discussion—in reality a gravitational wave has a very small amplitude (as formulated in linearized gravity). However they enable us to visualize the kind of oscillations associated with gravitational waves as produced for example by a pair of masses in a circular orbit. In this case the amplitude of the gravitational wave is a constant, but its plane of polarization changes or rotates at twice the orbital rate and so the time-varying gravitational wave size (or 'periodic spacetime strain') exhibits a variation as shown in the animation. If the orbit is elliptical then the gravitational wave's amplitude also varies with time according Einstein's quadrupole formula.

Like other waves, there are a few useful characteristics describing a gravitational wave:

- Amplitude: Usually denoted h, this is the size of the wave — the fraction of stretching or squeezing in the animation. The amplitude shown here is roughly $h = 0.5$ (or 50%). Gravitational waves passing through the Earth are many billions times weaker than this — $h \approx 10^{-20}$. Note that this is not the quantity that would be analogous to what is usually called the amplitude of an electromagnetic wave, which would be $\frac{\mathrm{d}h}{\mathrm{d}t}$.
- Frequency: Usually denoted f, this is the frequency with which the wave oscillates (1 divided by the amount of time between two successive maximum stretches or squeezes)
- Wavelength: Usually denoted λ, this is the distance along the wave between points of maximum stretch or squeeze.
- Speed: This is the speed at which a point on the wave (for example, a point of maximum stretch or squeeze) travels. For gravitational waves with small amplitudes, this is equal to the speed of light, c.

The speed, wavelength, and frequency of a gravitational wave are related by the equation $c = \lambda f$, just like the equation for a light wave. For example, the animations shown here oscillate roughly once every two seconds. This would correspond to a frequency of 0.5 Hz, and a wavelength of about 600,000 km, or 47 times the diameter of the Earth.

In the example just discussed, we actually assume something special about the wave. We have assumed that the wave is linearly polarized, with a "plus" polarization, written h_+. Polarization of a gravitational wave is just like polarization of a light wave except that the polarizations of a gravitational wave are at 45 degrees, as opposed to 90 degrees. In particular, if we had a "cross"-polarized gravitational wave, $h_\times$, the effect on the test particles would be basically the same, but rotated by 45 degrees, as shown in the second animation. Just as with light polarization, the polarizations of gravitational waves may also be expressed in terms of circularly polarized waves. Gravitational waves are polarized because of the nature of their sources. The polarization of a wave depends on the angle from the source, as we will see in the next section.

Sources of Gravitational Waves

In general terms, gravitational waves are radiated by objects whose motion involves acceleration, provided that the motion is not perfectly spherically symmetric (like an expanding or contracting sphere) or cylindrically symmetric (like a spinning disk or sphere). A simple example of this principle is provided by the spinning dumbbell. If the dumbbell spins like wheels on an axle, it will not radiate gravitational waves; if it tumbles end over end like two planets orbiting each other, it will radiate gravitational waves. The heavier the dumbbell, and the faster it tumbles, the greater is the gravitational radiation it will give off. If we imagine an extreme case in which the two weights of the dumbbell are massive stars like neutron stars or black holes, orbiting each other quickly, then significant amounts of gravitational radiation would be given off.

Some More Detailed Examples

- Two objects orbiting each other in a quasi-Keplerian planar orbit (basically, as a planet would orbit the Sun) *will* radiate.
- A spinning non-axisymmetric planetoid — say with a large bump or dimple on the equator — *will* radiate.
- A supernova *will* radiate except in the unlikely event that the explosion is perfectly symmetric.

- An isolated non-spinning solid object moving at a constant velocity *will not* radiate. This can be regarded as a consequence of the principle of conservation of linear momentum.
- A spinning disk *will not* radiate. This can be regarded as a consequence of the principle of conservation of angular momentum. However, it *will* show gravitomagnetic effects.
- A spherically pulsating spherical star (non-zero monopole moment or mass, but zero quadrupole moment) *will not* radiate, in agreement with Birkhoff's theorem.

More technically, the third time derivative of the quadrupole moment (or the *l*-th time derivative of the *l*-th multipole moment) of an isolated system's stress–energy tensor must be nonzero in order for it to emit gravitational radiation. This is analogous to the changing dipole moment of charge or current necessary for electromagnetic radiation.

Power Radiated by Orbiting Bodies

Gravitational waves carry energy away from their sources and, in the case of orbiting bodies, this is associated with an inspiral or decrease in orbit. Imagine for example a simple system of two masses — such as the Earth-Sun system — moving slowly compared to the speed of light in circular orbits. Assume that these two masses orbit each other in a circular orbit in the *x*-*y* plane. To a good approximation, the masses follow simple Keplerian orbits. However, such an orbit represents a changing quadrupole moment. That is, the system will give off gravitational waves.

Suppose that the two masses are m_1 and m_2, and they are separated by a distance r. The power given off (radiated) by this system is:

$$P = \frac{\mathrm{d}E}{\mathrm{d}t} = -\frac{32}{5}\frac{G^4}{c^5}\frac{(m_1 m_2)^2 (m_1 + m_2)}{r^5},$$

where G is the gravitational constant, c is the speed of light in vacuum and where the negative sign means that power is being given off by the system, rather than received. For a system like the Sun and Earth, r is about 1.5×10^{11} m and m_1 and m_2 are about 2×10^{30} and 6×10^{24} kg respectively. In this case, the power is about 200 watts. This is truly tiny compared to the total electromagnetic radiation given off by the Sun (roughly 3.86×10^{26} watts).

In theory, the loss of energy through gravitational radiation could eventually drop the Earth into the Sun. However, the total energy of the Earth orbiting the Sun (kinetic energy + gravitational potential energy) is about 1.14×10^{36} joules of which only 200 joules per second is lost through gravitational radiation, leading to a decay in the orbit by about 1×10^{-15} metres per day or roughly the diameter of a proton. At this rate, it would take the Earth approximately 1×10^{13} times more than the current age of the Universe to spiral onto the Sun. This estimate overlooks the decrease in r over time, but the majority of the time the bodies are far apart and only radiating slowly, so the difference is unimportant in this example.

A more dramatic example of radiated gravitational energy is represented by two solar mass neutron stars orbiting at a distance from each other of 1.89×10^{8} m (only 0.63 light-seconds apart). [The Sun is 8 light minutes from the Earth.] Plugging their masses into the above equation shows that the gravitational radiation from them would be 1.38×10^{28} watts, which is about 100 times more than the Sun's electromagnetic radiation.

Orbital Decay from Gravitational Radiation

Gravitational radiation robs the orbiting bodies of energy. It first circularizes their orbits and then gradually shrinks their radius. As the energy of the orbit is reduced, the distance between the bodies decreases, and they rotate more rapidly. The overall angular momentum is reduced however. This reduction corresponds to the angular momentum carried off by gravitational radiation. The rate of decrease of distance between the bodies versus time is given by:

$$\frac{dr}{dt} = -\frac{64}{5}\frac{G^3}{c^5}\frac{(m_1 m_2)(m_1 + m_2)}{r^3},$$

where the variables are the same as in the previous equation.

The orbit decays at a rate proportional to the inverse third power of the radius. When the radius has shrunk to half its initial value, it is shrinking eight times faster than before. By Kepler's Third Law, the new rotation rate at this point will be faster by $\sqrt{8} = 2.828$, or nearly three times the previous orbital frequency. As the radius decreases, the power lost to gravitational radiation increases even more. As can be seen from the previous equation, power radiated varies as the inverse fifth power of the radius, or 32 times more in this case.

If we use the previous values for the Sun and the Earth, we find that the Earth's orbit shrinks by 1.1×10^{-20} metre per second. This is 3.5×10^{-13} m per year, which is about 1/300 the diameter of a hydrogen atom. The effect of gravitational radiation on the size of the Earth's orbit is negligible over the age of the universe. This is not true for closer orbits.

A more practical example is the orbit of a Sun-like star around a heavy black hole. Our Milky Way has a, potential, 4 million solar-mass black hole at its centre in Sagittarius A. Such supermassive black holes are being found in the centre of almost all galaxies. For this example take a 2 million solar-mass black hole with a solar-mass star orbiting it at a radius of 1.89×10^{10} m (63 light-seconds). The mass of the black hole will be 4×10^{36} kg and its gravitational radius will be 6×10^{9} m. The orbital period will be 1,000 seconds, or a little under 17 minutes. The solar-mass star will draw closer to the black hole by 7.4 metres per second or 7.4 km per orbit. A collision will not be long in coming.

Assume that a pair of solar-mass neutron stars are in circular orbits at a distance of 1.89×10^{8} m (189,000 km). This is a little less than 1/7 the diameter of the Sun or 0.63 light-seconds. Their orbital period would be 1,000 seconds. Substituting the new mass and radius in the above formula gives a rate of orbit decrease of 3.7×10^{-6} m/s or 3.7 mm per orbit. This is 116 metres per year and is not negligible over cosmic time scales.

Suppose instead that these two neutron stars were orbiting at a distance of 1.89×10^{6} m (1890 km). Their period would be 1 second and their orbital velocity would be about 1/50 of the speed of light. Their orbit would now shrink by 3.7 metres per orbit. A collision is imminent. A runaway loss of energy from the orbit results in an ever more rapid decrease in the distance between the stars. They will eventually merge to form a black hole and cease to radiate gravitational waves. This is referred to as the inspiral.

The above equation can not be applied directly for calculating the lifetime of the orbit, because the rate of change in radius depends on the radius itself, and is thus non-constant with time. The lifetime can be computed by integration of this equation.

Orbital Lifetime Limits from Gravitational Radiation

Orbital lifetime is one of the most important properties of gravitational radiation sources. It determines the average number of binary stars in the universe that are close enough to be detected. Short lifetime binaries are strong sources of gravitational radiation but are few in number. Long lifetime binaries are more plentiful but they are weak sources of gravitational waves. LIGO is most sensitive in the frequency band where two neutron stars are about to merge. This time frame is only a few seconds. It takes luck for the detector to see this blink in time out of a million year orbital lifetime. It is predicted that such a merger will only be seen once per decade or so.

The lifetime of an orbit is given by:

$$t = \frac{5}{256}\frac{c^5}{G^3}\frac{r^4}{(m_1 m_2)(m_1 + m_2)},$$

where r is the initial distance between the orbiting bodies. This equation can be derived by integrating the previous equation for the rate of radius decrease. It predicts the time for the radius of the orbit to shrink to zero. As the orbital speed becomes a significant fraction of the speed of light, this equation becomes inaccurate. It is useful for inspirals until the last few milliseconds before the merger of the objects.

Substituting the values for the mass of the Sun and Earth as well as the orbital radius gives a very large lifetime of 3.44×10^{30} seconds or 1.09×10^{23} years (that is approximately 10^{13} times larger than the age of the universe). The actual figure would be slightly less than that. The Earth will break apart from tidal forces if it orbits closer than a few radii from the Sun. This would form a ring around the Sun and instantly stop the emission of gravitational waves.

If we use a 2 million solar mass black hole with a solar mass star orbiting it at 1.89×10^{10} metres, we get a lifetime of 6.50×10^{8} seconds or 20.7 years.

Assume that a pair of solar mass neutron stars with a diameter of 10 kilometres are in circular orbits at a distance of 1.89×10^{8} m (189,000 km). Their lifetime is 1.30×10^{13} seconds or about 414,000 years. Their orbital period will be 1,000 seconds and it could be observed by LISA if they were not too far away. A far greater number of white dwarf binaries exist with orbital periods in this range. White dwarf binaries have masses on the order of our Sun and diameters on the order of our Earth. They cannot get much closer together than 10,000 km before they will merge and cease to radiate gravitational waves. This results in the creation of either a neutron star or a black hole. Until then, their gravitational radiation will be comparable to that of a neutron star binary. LISA is the only gravitational wave experiment that is likely to succeed in detecting such types of binaries.

If the orbit of a neutron star binary has decayed to 1.89×10^{6}m (1890 km), its remaining lifetime is 130,000 seconds or about 36 hours. The orbital frequency will vary from 1 revolution per second at the start and 918 revolutions per second when the orbit has shrunk to 20 km at merger. The gravitational radiation emitted will be at twice the orbital frequency. Just before merger, the inspiral can be observed by LIGO if the binary is close enough. LIGO has only a few minutes to observe this merger out of a total orbital lifetime that may have been billions of years. The chance of success with LIGO as initially constructed is quite low despite the large number of such mergers occurring in the universe, because the sensitivity of the instrument does not 'reach' out to enough systems to see events frequently. No mergers have been seen in the few years that initial LIGO has been in operation, and it is thought that a merger should be seen about once per several tens of years of observing time with initial LIGO.

The upgraded Advanced LIGO detector, with a ten times greater sensitivity, 'reaches' out 10 times further—encompassing a volume 1000 times greater, and seeing 1000 times as many candidate sources. Thus, the expectation is that detections will be made at the rate of tens per year.

Wave Amplitudes from the Earth–Sun System

We can also think in terms of the amplitude of the wave from a system in circular orbits. Let θ be the angle between the perpendicular to the plane of the orbit and the line of sight of the observer. Suppose that an observer is outside the system at a distance R from its centre of mass. If R is much greater than a wavelength, the two polarizations of the wave will be

$$h_{+} = -\frac{1}{R}\frac{G^2}{c^4}\frac{2m_1m_2}{r}(1+\cos^2\theta)\cos\left[2\omega(t-R)\right],$$

$$h_{\times} = -\frac{1}{R}\frac{G^2}{c^4}\frac{4m_1m_2}{r}(\cos\theta)\sin\left[2\omega(t-R)\right].$$

Here, we use the constant angular velocity of a circular orbit in Newtonian physics:

$$\omega = \sqrt{G(m_1+m_2)/r^3}.$$

For example, if the observer is in the *x-y* plane then $\theta = \pi/2$, and $\cos(\theta) = 0$, so the $h_{\times}$ polarization is always zero. We also see that the frequency of the wave given off is twice the rotation frequency. If we put in numbers for the Earth-Sun system, we find:

$$h_{+} = -\frac{1}{R}\frac{G^2}{c^4}\frac{4m_1m_2}{r} = -\frac{1}{R}1.7\times10^{-10}\,\text{m}.$$

In this case, the minimum distance to find waves is $R \approx 1$ light-year, so typical amplitudes will be $h \approx 10^{-26}$. That is, a ring of particles would stretch or squeeze by just one part in 10^{26}. This is well under the detectability limit of all conceivable detectors.

Radiation from Other Sources

Although the waves from the Earth-Sun system are minuscule, astronomers can point to other sources for which the radiation should be substantial. One important example is the Hulse-Taylor binary — a pair of stars, one of which is a pulsar. The characteristics of their orbit can be deduced from the Doppler shifting of radio signals given off by the pulsar. Each of the stars has a mass about 1.4 times that of the Sun and the size of their orbit is about 1/75 of the Earth-Sun orbit. This means the distance between the two stars is just a few times larger than the diameter of our own Sun. The combination of greater masses and smaller separation means that the energy given off by the Hulse-Taylor binary will be far greater than the energy given off by the Earth-Sun system — roughly 10^{22} times as much.

The information about the orbit can be used to predict just how much energy (and angular momentum) should be given off in the form of gravitational waves. As the energy is carried off, the stars should draw closer to each other. This effect is called an inspiral, and it can be observed in the pulsar's signals. The measurements on the Hulse-Taylor system have been carried out over more than 30 years. It has been shown that the gravitational radiation predicted by general relativity allows these observations to be matched within 0.2 percent. In 1993, Russell Hulse and Joe Taylor were awarded the Nobel Prize in Physics for this work, which was the first indirect evidence for gravitational waves. Unfortunately, the orbital lifetime of this binary system before merger is about 1.84 billion years. This is a substantial fraction of the age of the universe.

Inspirals are very important sources of gravitational waves. Any time two compact objects (white dwarfs, neutron stars, or black holes) are in close orbits, they send out intense gravitational waves. As they spiral closer to each other, these waves become more intense. At some point they should become so intense that direct detection by their effect on objects on Earth or in space is possible. This direct detection is the goal of several large scale experiments.

The only difficulty is that most systems like the Hulse-Taylor binary are so far away. The amplitude of waves given off by the Hulse-Taylor binary as seen on Earth would be roughly $h \approx 10^{-26}$. There are some sources, however, that astrophysicists expect to find with much larger amplitudes of $h \approx 10^{-20}$. At least eight other binary pulsars have been discovered.

Astrophysics and Gravitational Waves

During the past century, astronomy has been revolutionized by the use of new methods for observing the universe. Astronomical observations were originally made using visible light. Galileo Galilei pioneered the use of telescopes to enhance these observations. However, visible light is only a small portion of the electromagnetic spectrum, and not all objects in the distant universe shine strongly in this particular band. More useful information may be found, for example, in radio wavelengths. Using radio telescopes, astronomers have found pulsars, quasars, and other extreme objects that push the limits of our understanding of physics. Observations in the microwave band have opened our eyes to the faint imprints of the Big Bang, a discovery Stephen Hawking called the "greatest discovery of the century, if not all time". Similar advances in observations using gamma rays, x-rays, ultraviolet light, and infrared light have also brought new insights to astronomy. As each of these regions of the spectrum has opened, new discoveries have been made that could not have been made otherwise. Astronomers hope that the same holds true of gravitational waves.

Gravitational waves have two important and unique properties. First, there is no need for any type of matter to be present nearby in order for the waves to be generated by a binary system of uncharged black holes, which would emit no electromagnetic radiation. Second, gravitational waves can pass through any intervening matter without being scattered significantly. Whereas light from distant stars may be blocked out by interstellar dust, for example, gravitational waves will pass through essentially unimpeded. These two features allow gravitational waves to carry information about astronomical phenomena never before observed by humans.

The sources of gravitational waves described above are in the low-frequency end of the gravitational-wave spectrum (10^{-7} to 10^{5} Hz). An astrophysical source at the high-frequency end of the gravitational-wave spectrum (above 10^{5} Hz and probably 10^{10} Hz) generates relic gravitational waves that are theorized to be faint imprints of the Big Bang like the cosmic microwave background. At these high frequencies it is potentially possible that the sources may be "man made" that is, gravitational waves generated and detected in the laboratory.

Gravitational Wave Background

A possible target of gravitational wave detection experiments is a stochastic background of gravitational waves. This background is known as the gravitational wave background or the stochastic background. The detection of such a background would have a profound impact on early universe cosmology and on high-energy physics, opening up a new window and exploring very early times in the evolution of the universe, and correspondingly high energies, that will never be accessible by other means.

The emission of gravitational waves from a large number of unresolved astrophysical sources can create a stochastic background of gravitational waves. For instance, sufficiently massive stars, at the final stage of their evolution, collapse to form a black hole or a neutron star. In this explosive supernova event gravitational waves are liberated. Also, in rapidly rotating neutron stars there is a whole class of instabilities driven by the emission of gravitational waves.

A stochastic gravitational wave background is also of theoretical interest. Given its stochastic nature, a coincidence of gravitational waves at a particular point could create stress-energy densities sufficient to produce an event horizon. This would produce a relativistic explanation of nonlocal effects.

Energy, Momentum, and Angular Momentum Carried by Gravitational Waves

Waves familiar from other areas of physics such as water waves, sound waves, and electromagnetic waves are able to carry energy, momentum, and angular momentum. By carrying these away from a source, waves are able to rob that source of its energy as well as its linear and angular momentum. Gravitational waves perform the same function. Thus, for example, a binary system loses angular momentum as the two orbiting objects spiral towards each other—the angular momentum is radiated away by gravitational waves.

The waves can also carry off linear momentum, a possibility that has some interesting implications for astrophysics. After two supermassive black holes coalesce, emission of linear momentum can produce a “kick” with amplitude as large as 4000 km/s. This is fast enough to eject the coalesced black hole completely from its host galaxy. Even if the kick is too small to eject the black hole completely, it can remove it temporarily from the nucleus of the galaxy, after which it will oscillate about the centre, eventually coming to rest. A kicked black hole can also carry a star cluster with it, forming a hyper-compact stellar system. Or it may carry gas, allowing the recoiling black hole to appear temporarily as a “naked quasar”. The quasar SDSS J092712.65+294344.0 is believed to contain a recoiling supermassive black hole.

Detecting Gravitational Waves

Difficulties in Detection

Figure: *Evidence of gravitational waves in the infant universe may have been uncovered by the BICEP2 radio telescope. The microscopic examination of the focal plane of the BICEP2 detector is shown here.*

Gravitational waves are not easily detectable. This knowledge gap is primarily due to the massive presence of noise in the low frequencies where antennas currently operate. Gravitational waves are expected to have frequencies $10^{-16}\,\text{Hz} < f < 10^{4}\,\text{Hz}$.

Gravitational Wave Detector

A gravitational-wave observatory (or gravitational-wave detector) is any device designed to measure gravitational waves, tiny distortions of spacetime that were first predicted by Einstein in 1916. Gravitational waves are perturbations in the curvature of spacetime caused by accelerated masses.

The existence of gravitational radiation is a specific prediction of general relativity, but is a feature of all theories of gravity that obey special relativity. Since the 1960s, gravitational-wave detectors have been built and constantly improved. The present-day generation of resonant mass antennas and laser interferometres has reached the necessary sensitivity to detect gravitational waves from sources in the Milky Way. Gravitational-wave observatories are the primary tool of gravitational-wave astronomy.

The direct detection of gravitational waves is complicated by the extraordinarily small effect the waves would produce on a detector. The amplitude of a spherical wave will fall off as the inverse of the distance from the source. Thus, even waves from extreme systems like merging binary black holes die out to very small amplitude by the time they reach the Earth. Astrophysicists expect that some gravitational waves passing the Earth may be as large as $h \approx 10^{-18}$, but generally no bigger.

Weber Bars

A simple device to detect the expected wave motion is called a Weber bar – a large, solid bar of metal isolated from outside vibrations. This type of instrument was the first type of gravitational-wave detector. Strains in space due to an incident gravitational wave excite the bar's resonant frequency and could thus be amplified to detectable levels. Conceivably, a nearby supernova might be strong enough to be seen without resonant amplification. Modern forms of the Weber bar are still operated, cryogenically cooled, with superconducting quantum interference devices to detect vibration. Weber bars are not sensitive enough to detect anything but extremely powerful gravitational waves.

MiniGRAIL is a spherical gravitational-wave antenna using this principle. It is based at Leiden University, consisting of an exactingly machined 1150 kg sphere cryogenically cooled to 20 mK. The spherical configuration allows for equal sensitivity in all directions, and is somewhat experimentally simpler than larger linear devices requiring high vacuum. Events are detected by measuring deformation of the detector sphere. MiniGRAIL is highly sensitive in the 2–4 kHz range, suitable for detecting gravitational waves from rotating neutron star instabilities or small black hole mergers.

AURIGA is an ultracryogenic resonant bar gravitational wave detector based at INFN in Italy. It is based on a cylindrical bar detector. The AURIGA and LIGO teams have collaborated in joint observations.

Interferometres

A more sensitive detector uses laser interferometry to measure gravitational-wave induced motion between separated 'free' masses. This allows the masses to be separated by large distances (increasing the signal size); a further advantage is that it is sensitive to a wide range of frequencies (not just those near a resonance as is the case for Weber bars). Ground-based interferometres are now operational. Currently, the most sensitive is LIGO – the Laser Interferometre Gravitational Wave Observatory. LIGO has three detectors: one in

Livingston, Louisiana; the other two (in the same vacuum tubes) at the Hanford site in Richland, Washington. Each consists of two light storage arms which are 2 to 4 kilometres in length. These are at 90 degree angles to each other, with the light passing through 1m diameter vacuum tubes running the entire 4 kilometres. A passing gravitational wave will slightly stretch one arm as it shortens the other. This is precisely the motion to which an interferometre is most sensitive.

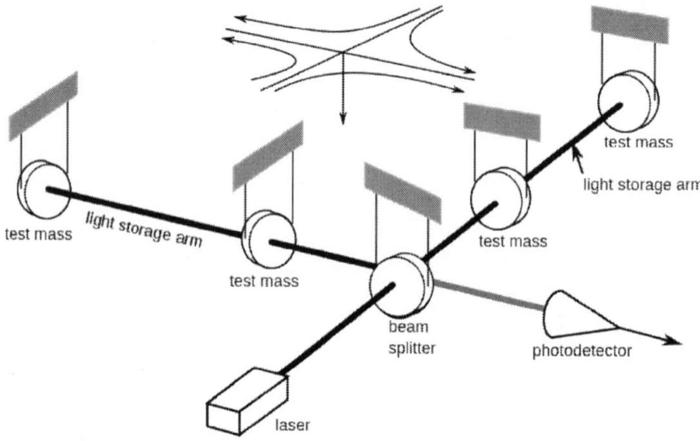

Figure: *A schematic diagram of a laser interferometre*

Even with such long arms, the strongest gravitational waves will only change the distance between the ends of the arms by at most roughly 10^{-18} metres. LIGO should be able to detect gravitational waves as small as $h \approx 5 \times 10^{-22}$. Upgrades to LIGO and other detectors such as VIRGO, GEO 600, and TAMA 300 should increase the sensitivity still further; the next generation of instruments (Advanced LIGO and Advanced Virgo) will be more than ten times more sensitive. Another highly sensitive interferometre (LCGT) is currently in the design phase. A key point is that a ten-times increase in sensitivity (radius of "reach") increases the volume of space accessible to the instrument by one thousand. This increases the rate at which detectable signals should be seen from one per tens of years of observation, to tens per year.

Interferometric detectors are limited at high frequencies by shot noise, which occurs because the lasers produce photons randomly; one analogy is to rainfall – the rate of rainfall, like the laser intensity, is measurable, but the raindrops, like photons, fall at random times, causing fluctuations around the average value. This leads to noise at the output of the detector, much like radio static. In addition, for sufficiently high laser power, the random momentum transferred to the test masses by the laser photons shakes the mirrors, masking signals at low frequencies. Thermal noise (e.g., Brownian motion) is another limit to sensitivity. In addition to these "stationary" (constant) noise sources, all ground-based detectors are also limited at low frequencies by seismic noise and other forms of environmental vibration, and other "non-stationary" noise sources; creaks in mechanical structures, lightning or other large electrical disturbances, etc. may also create noise masking an event or may even imitate an event. All these must be taken into account and excluded by analysis before a detection may be considered a true gravitational-wave event.

Space-based interferometres, such as LISA and DECIGO, are also being developed. LISA's design calls for three test masses forming an equilateral triangle, with lasers from each spacecraft to each other spacecraft forming two independent interferometres. LISA is planned to occupy a solar orbit trailing the Earth, with each arm of the triangle being five million kilometres. This puts the detector in an excellent vacuum far from Earth-based sources of noise, though it will still be susceptible to shot noise, as well as artifacts caused by cosmic rays and solar wind.

An *atomic gravitational wave interferometric sensor* (AGIS) is a novel detection scheme to detect gravitational waves, proposed by S. Dimopoulos et al. in 2008.

Ground-Based Interferometres

Though the Hulse-Taylor observations were very important, they give only *indirect* evidence for gravitational waves. A more conclusive observation would be a *direct* measurement of the effect of a passing gravitational wave, which could also provide more information about the system that generated it. Any such direct detection is complicated by the extraordinarily small effect the waves would produce on a detector. The amplitude of a spherical wave will fall off as the inverse of the distance from the source (the $1/R$ term in the formulas for h above). Thus, even waves from extreme systems like merging binary black holes die out to very small amplitude by the time they reach the Earth. Astrophysicists expect that some gravitational waves passing the Earth may be as large as $h \approx 10^{-20}$, but generally no bigger.

A simple device theorised to detect the expected wave motion is called a Weber bar — a large, solid bar of metal isolated from outside vibrations. This type of instrument was the first type of gravitational wave detector. Strains in space due to an incident gravitational wave excite the bar's resonant frequency and could thus be amplified to detectable levels. Conceivably, a nearby supernova might be strong enough to be seen without resonant amplification. With this instrument, Joseph Weber claimed to have detected daily signals of gravitational waves. His results, however, were contested in 1974 by physicists Richard Garwin and David Douglass. Modern forms of the Weber bar are still operated, cryogenically cooled, with superconducting quantum interference devices to detect vibration. Weber bars are not sensitive enough to detect anything but extremely powerful gravitational waves.

MiniGRAIL is a spherical gravitational wave antenna using this principle. It is based at Leiden University, consisting of an exactingly machined 1150 kg sphere cryogenically cooled to 20 mK. The spherical configuration allows for equal sensitivity in all directions, and is somewhat experimentally simpler than larger linear devices requiring high vacuum. Events are detected by measuring deformation of the detector sphere. MiniGRAIL is highly sensitive in the 2–4 kHz range, suitable for detecting gravitational waves from rotating neutron star instabilities or small black hole mergers.

A more sensitive class of detector uses laser interferometry to measure gravitational-wave induced motion between separated 'free' masses. This allows the masses to be separated by large distances (increasing the signal size); a further advantage is that it is sensitive to a wide range of frequencies (not just those near a resonance as is the case for Weber bars). Ground-based interferometres are now operational. Currently, the most sensitive is LIGO — the Laser Interferometre Gravitational Wave Observatory. LIGO has three detectors: one in Livingston, Louisiana; the other two (in the same vacuum tubes) at the Hanford site in Richland, Washington. Each consists of two light storage arms that are 2 to 4 kilometres in length. These are at 90 degree angles to each other, with the light passing through 1m diameter vacuum tubes running the entire 4 kilometres. A passing gravitational wave will slightly stretch one arm as it shortens the other. This is precisely the motion to which an interferometre is most sensitive.

Even with such long arms, the strongest gravitational waves will only change the distance between the ends of the arms by at most roughly 10^{-18} metres. LIGO should be able to detect gravitational waves as small as $h \sim 5 \times 10^{-20}$. Upgrades to LIGO and other detectors such as Virgo, GEO 600, and TAMA 300 should increase the sensitivity still further; the next generation of instruments (Advanced LIGO and Advanced Virgo) will be more than ten times more sensitive. Another highly sensitive interferometre (LCGT) is currently in the design phase. A key point is that a tenfold increase in sensitivity (radius of 'reach') increases the volume of space accessible to the instrument by one thousand times. This increases the rate at which detectable signals should be seen from one per tens of years of observation, to tens per year.

Interferometric detectors are limited at high frequencies by shot noise, which occurs because the lasers produce photons randomly; one analogy is to rainfall—the rate of rainfall, like the laser intensity, is measurable, but the raindrops, like photons, fall at random times, causing fluctuations around the average value. This leads to noise at the output of the detector, much like radio static. In addition, for sufficiently high laser power, the random momentum transferred to the test masses by the laser photons shakes the mirrors, masking signals at low frequencies. Thermal noise (e.g., Brownian motion) is another limit to sensitivity. In addition to these 'stationary' (constant) noise sources, all ground-based detectors are also limited at low frequencies by seismic noise and other forms of environmental vibration, and other 'non-stationary' noise sources; creaks in mechanical structures, lightning or other large electrical disturbances, etc. may also create noise masking an event or may even imitate an event. All these must be taken into account and excluded by analysis before a detection may be considered a true gravitational wave event.

Space-based interferometres, such as LISA and DECIGO, are also being developed. LISA's design calls for three test masses forming an equilateral triangle, with lasers from each spacecraft to each other spacecraft forming two independent interferometres. LISA is planned to occupy a solar orbit trailing the Earth, with each arm of the triangle being five million kilometres. This puts the detector in an excellent vacuum far from Earth-based sources of noise, though it will still be susceptible to shot noise, as well as artifacts caused by cosmic rays and solar wind.

There are currently two detectors focusing on detection at the higher end of the gravitational wave spectrum (10^{-7} to 10^{5} Hz): one at University of Birmingham, England, and the other at INFN Genoa, Italy. A third is under development at Chongqing University, China. The Birmingham detector measures changes in the polarization state of a microwave beam circulating in a closed loop about one metre across. Two have been fabricated and they are currently expected to be sensitive to periodic spacetime strains of $h \sim 2 \times 10^{-13} / \sqrt{\text{Hz}}$, given as an amplitude spectral density. The INFN Genoa detector is a resonant antenna consisting of two coupled spherical superconducting harmonic oscillators a few centimetres in diameter. The oscillators are designed to have (when uncoupled) almost equal resonant frequencies. The system is currently expected to have a sensitivity to periodic spacetime strains of $h \sim 2 \times 10^{-17} / \sqrt{\text{Hz}}$, with an expectation to reach a sensitivity of $h \sim 2 \times 10^{-20} / \sqrt{\text{Hz}}$. The Chongqing University detector is planned to detect relic high-frequency gravitational waves with the predicted typical parameters $?_g \sim 10^{10}$ Hz (10 GHz) and $h \sim 10^{-30}$-10^{-31}.

Using Pulsar Timing Arrays

Pulsars are rapidly rotating stars. A pulsar emits beams of radio waves that, like lighthouse beams, sweep through the sky as the pulsar rotates. The signal from a pulsar can be detected by radio telescopes as a series of regularly spaced pulses, essentially like the ticks of a clock. Gravitational waves affect the time it takes the pulses to travel from the pulsar to a telescope on Earth.

A pulsar timing array uses millisecond pulsars to seek out perturbations due to gravitational waves in measurements of pulse arrival times at a telescope, in other words, to look for deviations in the clock ticks. In particular, pulsar timing arrays can search for a distinct pattern of correlation and anti-correlation between the signals over an array of different pulsars (resulting in the name "pulsar timing array"). Although pulsar pulses travel through space for hundreds or thousands of years to reach us, pulsar timing arrays are sensitive to perturbations in their travel time of much less than a millionth of a second.

Globally there are three active pulsar timing array projects. The North American Nanohertz Gravitational Wave Observatory uses data collected by the Arecibo Radio Telescope and Green Bank Telescope. The Parkes Pulsar Timing Array at the Parkes radio-telescope has been collecting data since March 2005. The European Pulsar Timing Array uses data from the four largest telescopes in Europe: the Lovell Telescope, the Westerbork Synthesis Radio Telescope, the Effelsberg Telescope and the Nancay Radio Telescope. (Upon completion the Sardinia Radio Telescope will be added to the EPTA also.) These three projects have begun collaborating under the title of the International Pulsar Timing Array project.

Einstein@Home

In some sense, the easiest signals to detect should be constant sources. Supernovae and neutron star or black hole mergers should have larger amplitudes and be more interesting, but the waves generated will be more complicated. The waves given off by a spinning, aspherical neutron star would be "monochromatic"—like a pure tone in acoustics. It would not change very much in amplitude or frequency.

The Einstein@Home project is a distributed computing project similar to SETI@home intended to detect this type of simple gravitational wave. By taking data from LIGO and GEO, and sending it out in little pieces to thousands of volunteers for parallel analysis on their home computers, Einstein@Home can sift through the data far more quickly than would be possible otherwise.

Primordial Gravitational Waves

Primordial gravitational waves are gravitational waves observed in the cosmic microwave background.

Reports stated in March 2014 that BICEP instrument have detected *B*-modes from gravitational waves in the early universe. An announcement was made on 17 March 2014 from the Harvard–Smithsonian Centre for Astrophysics announcing that BICEP2 had detected B-modes at the level of $r = 0.20{+}0.07{-}0.05$, disfavouring the null hypothesis ($r = 0$) at the level of 7 sigma (5.9σ after foreground subtraction).

Models of cosmic inflation predict that gravitational waves should appear; thus, their detection supports the theory of inflation, and the strength can confirm and exclude different models of inflation.

However, on 19 June 2014, lowered confidence in confirming the findings was reported and on 19 September 2014 new results of the Planck experiment reported that the results of BICEP2 can be fully attributed to dust. Hence, there is no experimental evidence for primordial gravitational waves, to date.

Mathematics

Einstein's equations form the fundamental law of general relativity. The curvature of spacetime can be expressed mathematically using the metric tensor — denoted $g_{\mu\nu}$. The metric holds information regarding how distances are measured in the space under consideration. Because the propagation of gravitational waves through space and time change distances, we will need to use this to find the solution to the wave equation.

Spacetime curvature is also expressed with respect to a covariant derivative, ∇ , in the form of the Einstein tensor, $G_{\mu\nu}$. This curvature is related to the stress–energy tensor, $T_{\mu\nu}$, by the key equation

$$G_{\mu\nu} = \frac{8\pi G_N}{c^4} T_{\mu\nu},$$

where G_N is Newton's gravitational constant, and c is the speed of light. We assume geometrized units, so $G_N = 1 = c$.

With some simple assumptions, Einstein's equations can be rewritten to show explicitly that they are wave equations. To begin with, we adopt some coordinate system, like (t, r, θ, ϕ). We define the "flat-space metric" $\eta_{\mu\nu}$ to be the quantity that — in this coordinate system — has the components we would expect for the flat space metric. For example, in these spherical coordinates, we have

$$\eta_{\mu\nu} = \begin{bmatrix} -1 & 0 & 0 & 0 \\ 0 & 1 & 0 & 0 \\ 0 & 0 & r^2 & 0 \\ 0 & 0 & 0 & r^2 \sin^2\theta \end{bmatrix}.$$

This flat-space metric has no physical significance; it is a purely mathematical device necessary for the analysis. Tensor indices are raised and lowered using this "flat-space metric".

Now, we can also think of the physical metric $g_{\mu\nu}$ as a matrix, and find its determinant, $\det g$. Finally, we define a quantity

$$\bar{h}^{\alpha\beta} \equiv \eta^{\alpha\beta} - \sqrt{|\det g|}\, g^{\alpha\beta} .$$

This is the crucial field, which will represent the radiation. It is possible (at least in an asymptotically flat spacetime) to choose the coordinates in such a way that this quantity satisfies the "de Donder" gauge conditions (conditions on the coordinates):

$$\nabla_\beta \bar{h}^{\alpha\beta} = 0,$$

where ∇ represents the flat-space derivative operator. These equations say that the divergence of the field is zero. The linear Einstein equations can now be written as

$$\Box \bar{h}^{\alpha\beta} = -16\pi\tau^{\alpha\beta} ,$$

where $\Box = -\partial_t^2 + \Delta$ represents the flat-space d'Alembertian operator, and represents the stress–energy tensor plus quadratic terms involving $\bar{h}^{\alpha\beta}$. This is just a wave equation for the field with a source, despite the fact that the source involves terms quadratic in the field itself. That is, it can be shown that solutions to this equation are waves travelling with velocity 1 in these coordinates.

Linear Approximation

The equations above are valid everywhere — near a black hole, for instance. However, because of the complicated source term, the solution is generally too difficult to find analytically.

We can often assume that space is nearly flat, so the metric is nearly equal to the $\eta^{\alpha\beta}$ tensor. In this case, we can neglect terms quadratic in $\bar{h}^{\alpha\beta}$, which means that the $\tau^{\alpha\beta}$ field reduces to the usual stress–energy tensor $T^{\alpha\beta}$. That is, Einstein's equations become

$$\Box \bar{h}^{\alpha\beta} = -16\pi T^{\alpha\beta} .$$

If we are interested in the field far from a source, however, we can treat the source as a point source; everywhere else, the stress–energy tensor would be zero, so

$$\Box \bar{h}^{\alpha\beta} = 0 .$$

Now, this is the usual homogeneous wave equation — one for each component of $\bar{h}^{\alpha\beta}$. Solutions to this equation are well known. For a wave moving away from a point source, the radiated part (meaning the part that dies off as $1/r$ far from the source) can always be written in the form $A(t-r,\theta,\phi)/r$, where A is just some function.

It can be shown that — to a linear approximation — it is always possible to make the field traceless.

Now, if we further assume that the source is positioned at $r=0$, the general solution to the wave equation in spherical coordinates is

$$\bar{h}^{\alpha\beta} = \frac{1}{r}\begin{bmatrix} 0 & 0 & 0 & 0 \\ 0 & 0 & 0 & 0 \\ 0 & 0 & A_{+}(t-r,\theta,\phi) & A_{\times}(t-r,\theta,\phi) \\ 0 & 0 & A_{\times}(t-r,\theta,\phi) & -A_{+}(t-r,\theta,\phi) \end{bmatrix}$$

$$\equiv \begin{bmatrix} 0 & 0 & 0 & 0 \\ 0 & 0 & 0 & 0 \\ 0 & 0 & h_{+}(t-r,r,\theta,\phi) & h_{\times}(t-r,r,\theta,\phi) \\ 0 & 0 & h_{\times}(t-r,r,\theta,\phi) & -h_{+}(t-r,r,\theta,\phi) \end{bmatrix}$$

where we now see the origin of the two polarizations.

Relation to the Source

If we know the details of a source — for instance, the parameters of the orbit of a binary — we can relate the source's motion to the gravitational radiation observed far away. With the relation

$$\Box \bar{h}^{\alpha\beta} = -16\pi\tau^{\alpha\beta} ,$$

we can write the solution in terms of the tensorial Green's function for the d'Alembertian operator:

$$\bar{h}^{\alpha\beta}(t,\vec{x}) = -16\pi \int G^{\alpha\beta}_{\gamma\delta}(t,\vec{x};t',\vec{x}')\tau^{\gamma\delta}(t',\vec{x}')\mathrm{d}t'\mathrm{d}^3x' .$$

Though it is possible to expand the Green's function in tensor spherical harmonics, it is easier to simply use the form

$$G^{\alpha\beta}_{\gamma\delta}(t,\vec{x};t',\vec{x}') = \frac{1}{4\pi}\delta^{\alpha}_{\gamma}\,\delta^{\beta}_{\delta}\frac{\delta(t\pm|\vec{x}-\vec{x}'|-t')}{|\vec{x}-\vec{x}'|},$$

where the positive and negative signs correspond to ingoing and outgoing solutions, respectively.

Generally, we are interested in the outgoing solutions, so

$$\bar{h}^{\alpha\beta}(t,\vec{x}) = -4\int \frac{\tau^{\alpha\beta}(t-|\vec{x}-\vec{x}'|,\vec{x}')}{|\vec{x}-\vec{x}'|}\mathrm{d}^3x' .$$

If the source is confined to a small region very far away, to an excellent approximation we have:

$$\bar{h}^{\alpha\beta}(t,\vec{x}) \approx -\frac{4}{r}\int \tau^{\alpha\beta}(t-r,\vec{x}')\mathrm{d}^3x' ,$$

where $r = |\vec{x}|$.

Now, because we will eventually only be interested in the spatial components of this equation (time components can be set to zero with a coordinate transformation), and we are integrating this quantity — presumably over a region of which there is no boundary — we can put this in a different form.

Ignoring divergences with the help of Stokes' theorem and an empty boundary, we can see that

$$\int \tau^{ij}(t-r,\vec{x}')\mathrm{d}^3x' = \int x'^i x'^j \nabla_k \nabla_l \tau^{kl}(t-r,\vec{x}')\mathrm{d}^3x' ,$$

Inserting this into the above equation, we arrive at

$$\bar{h}^{ij}(t,\vec{x}) \approx -\frac{4}{r}\int x'^i x'^j \nabla_k \nabla_l \tau^{kl}(t-r,\vec{x}')\mathrm{d}^3x' ,$$

Finally, because we have chosen to work in coordinates for which $\nabla_\beta \bar{h}^{\alpha\beta} = 0$, we know that $\nabla_\beta \tau^{\alpha\beta} = 0$. With a few simple manipulations, we can use this to prove that

$$\nabla_0 \nabla_0 \tau^{00} = \nabla_j \nabla_k \tau^{jk}.$$

With this relation, the expression for the radiated field is

$$\bar{h}^{ij}(t,\vec{x}) \approx -\frac{4}{r}\frac{\mathrm{d}^2}{\mathrm{d}t^2}\int x'^i x'^j \tau^{00}(t-r,\vec{x}')\mathrm{d}^3x'.$$

In the linear case, $\tau^{00} = \rho$, the density of mass-energy.

To a very good approximation, the density of a simple binary can be described by a pair of delta-functions, which eliminates the integral. Explicitly, if the masses of the two objects are M_1 and M_2, and the positions are $\vec{x}_1$ and $\vec{x}_2$, then

$$\rho(t-r,\vec{x}') = M_1\delta^3(\vec{x}' - \vec{x}_1(t-r)) + M_2\delta^3(\vec{x}' - \vec{x}_2(t-r)).$$

We can use this expression to do the integral above:

$$\bar{h}^{ij}(t,\vec{x}) \approx -\frac{4}{r}\frac{\mathrm{d}^2}{\mathrm{d}t^2}\left\{M_1 x_1^i(t-r)x_1^j(t-r) + M_2 x_2^i(t-r)x_2^j(t-r)\right\}.$$

Using mass-centred coordinates, and assuming a circular binary, this is

$$\bar{h}^{ij}(t,\vec{x}) \approx -\frac{4}{r}\frac{M_1 M_2}{R} n^i(t-r)n^j(t-r),$$

where $\vec{n} = \vec{x}_1/|\vec{x}_1|$. Plugging in the known values of $\vec{x}_1(t-r)$, we obtain the expressions given above for the radiation from a simple binary.

Gravity Wave

In fluid dynamics, gravity waves are waves generated in a fluid medium or at the interface between two media when the force of gravity or buoyancy tries to restore equilibrium. An example of such an interface is that between the atmosphere and the ocean, which gives rise to wind waves.

When a fluid element is displaced on an interface or internally to a region with a different density, gravity will try to restore it toward equilibrium, resulting in an oscillation about the equilibrium state or *wave orbit*. Gravity waves on an air–sea interface of the ocean are called surface gravity waves or surface waves, while gravity waves that are *within* the body of the water (such as between parts of different densities) are called internal waves. Wind-generated waves on the water surface are examples of gravity waves, and tsunamis and ocean tides are others.

Wind-generated gravity waves on the free surface of the Earth's ponds, lakes, seas and oceans have a period of between 0.3 and 30 seconds (3 Hz to 0.03 Hz). Shorter waves are also affected by surface tension and are called gravity–capillary waves and (if hardly influenced by gravity) capillary waves.

Alternatively, so-called infragravity waves, which are due to subharmonic nonlinear wave interaction with the wind waves, have periods longer than the accompanying wind-generated waves.

Atmosphere Dynamics on Earth

In the Earth's atmosphere, gravity waves are a mechanism for the transfer of momentum from the troposphere to the stratosphere. Gravity waves are generated in the troposphere by frontal systems or by airflow over mountains. At first, waves propagate through the atmosphere without appreciable change in mean velocity. But as the waves reach more rarefied air at higher altitudes, their amplitude increases, and nonlinear effects cause the waves to break, transferring their momentum to the mean flow.

This process plays a key role in studying the dynamics of the middle atmosphere.

The clouds in gravity waves can look like altostratus undulatus clouds, and are sometimes confused with them, but the formation mechanism is different

Quantitative Description

The phase velocity c of a linear gravity wave with wavenumber k is given by the formula

$$c = \sqrt{\frac{g}{k}},$$

where g is the acceleration due to gravity. When surface tension is important, this is modified to

$$c = \sqrt{\frac{g}{k} + \frac{\sigma k}{\rho}},$$

where σ is the surface tension coefficient and ρ is the density.

Details of the Phase-Speed Derivation

Since $c = \omega / k$ is the phase speed in terms of the angular frequency ω and the wavenumber, the gravity wave angular frequency can be expressed as

$$\omega = \sqrt{gk}.$$

The group velocity of a wave (that is, the speed at which a wave packet travels) is given by

$$c_g = \frac{d\omega}{dk},$$

and thus for a gravity wave,

$$c_g = \frac{1}{2}\sqrt{\frac{g}{k}} = \frac{1}{2}c.$$

The group velocity is one half the phase velocity. A wave in which the group and phase velocities differ is called dispersive.

The Generation of Waves by Wind

Wind waves, as their name suggests, are generated by wind transferring energy from the atmosphere to the ocean's surface, and capillary-gravity waves play an essential role in this effect. There are two distinct mechanisms involved, called after their proponents, Phillips and Miles.

In the work of Phillips, the ocean surface is imagined to be initially flat (*glassy*), and a turbulent wind blows over the surface. When a flow is turbulent, one observes a randomly fluctuating velocity field superimposed on a mean flow (contrast with a laminar flow, in which the fluid motion is ordered and smooth). The fluctuating velocity field gives rise to fluctuating stresses (both tangential and normal) that act on the air-water interface. The normal stress, or fluctuating pressure acts as a forcing term (much like pushing a swing introduces a forcing term). If the frequency and wavenumber (ω, k) of this forcing term match a mode of vibration of the capillary-gravity wave (as derived above), then there is a resonance, and the wave grows in amplitude. As with other resonance effects, the amplitude of this wave grows linearly with time.

The air-water interface is now endowed with a surface roughness due to the capillary-gravity waves, and a second phase of wave growth takes place. A wave established on the surface either spontaneously as described above, or in laboratory conditions, interacts with the turbulent mean flow in a manner described by Miles. This is the so-called critical-layer mechanism. A critical layer forms at a height where the wave speed c equals the mean turbulent flow U.

As the flow is turbulent, its mean profile is logarithmic, and its second derivative is thus negative. This is precisely the condition for the mean flow to impart its energy to the interface through the critical layer. This supply of energy to the interface is destabilizing and causes the amplitude of the wave on the interface to grow in time. As in other examples of linear instability, the growth rate of the disturbance in this phase is exponential in time.

This Miles–Phillips Mechanism process can continue until an equilibrium is reached, or until the wind stops transferring energy to the waves (i.e., blowing them along) or when they run out of ocean distance, also known as fetch length.

Undular Bore

In meteorology, an undular bore is a wave disturbance in the Earth's atmosphere and can be seen through unique cloud formations. They normally occur within an area of the atmosphere which is stable in the low levels after an outflow boundary or a cold front moves through.

In hydraulics, an undular bore is a gentle bore with an undular hydraulic jump pattern at the downstream (subcritical) side.

Undular bores are usually formed when two air masses of different temperatures collide. When a low level boundary such as a cold front or outflow boundary approaches a layer of cold, stable air, it creates a disturbance in the atmosphere producing a wave-like motion, known as a gravity wave.

Although the undular bore waves appear as bands of clouds across the sky, they are transverse waves, and are propelled by the transfer of energy from an oncoming storm and are shaped by gravity. The ripple like appearance of this wave is described as the disturbance in the water when a pebble is dropped into a pond or when a moving boat creates waves in the surrounding water. The object displaces the water or medium the wave is travelling through and the medium moves in an upward motion. However, because of gravity, the water or medium is pulled back down and the repetition of this cycle creates the transverse wave motion.

The undular bore's period can measure 5 miles (8.0 km) peak to peak and can travel 16 kilometres per hour (9.9 mph) to 95 kilometres per hour (59 mph). The medium it travels through is the atmosphere. There are several varying types of "bores" in different layers of the atmosphere, such as the mesospheric bore which occurs in the mesosphere.

Undular bores are believed to be catalysts for thunderstorms. Although a thunderstorm helps create an undular bore, an undular bore can in turn intensify a thunderstorm because it further disturbs the atmosphere. "Undular bores can have an effect on severe weather conditions," said Tim Coleman, Atmospheric Scientist, National Space Science and Technology Centre (NSSTC). "A tornado in Birmingham, Alabama in April 1998 that came in contact with an undular bore increased in size and intensity."

Occurrences

Rare but not unknown in a great many locations, the waves appear with some predictability and regularity in the Gulf of Carpentaria during Spring. They have been seen as frequently as six days in a row according to reports by the two pilots who have most experience with soaring these sometimes enormous examples of the undular bore, known in Australia as the Morning Glory cloud.

In Hydraulics

The term "bore" is also used to describe positive surges advancing in shallow waters. When the surge's Froude number is less than 1.4 to 1.7 (i.e. above unity and below a number somewhere in the range 1.4 to 1.7), the advancing front is followed by a train of well-defined free-surface undulations (called "*whelps*"). The surge is then called an undular surge or undular bore.

The undulations form a standing wave pattern, relative to the undular bore front. So, the phase velocity (propagation velocity relative to still water) of the undulations is just high enough to keep the undulations stationary relative to the bore front. Now, in water waves, the group velocity (which is also the energy-transport velocity) is less than the phase velocity. Therefore, on average, wave energy of the undulations is transported away from the front, and contributing to the energy loss in the region of the front.

A related occurrence of positive surges is the tidal bore in estuaries.

Lee Wave

In meteorology, lee waves are atmospheric standing waves. The most common form is mountain waves, which are atmospheric internal gravity waves. These were discovered in 1933 by two German glider pilots, Hans Deutschmann and Wolf Hirth, above the Krkonoše. They are periodic changes of atmospheric pressure, temperature and orthometric height in a current of air caused by vertical displacement, for example orographic lift when the wind blows over a mountain or mountain range. They can also be caused by the surface wind blowing over an escarpment or plateau, or even by upper winds deflected over a thermal updraft or cloud street.

The vertical motion forces periodic changes in speed and direction of the air within this air current. They always occur in groups on the lee side of the terrain that triggers them. Usually a turbulent vortex, with its axis of rotation parallel to the mountain range, is generated around the first trough; this is called a rotor. The strongest lee waves are produced when the lapse rate shows a stable layer above the obstruction, with an unstable layer above and below.

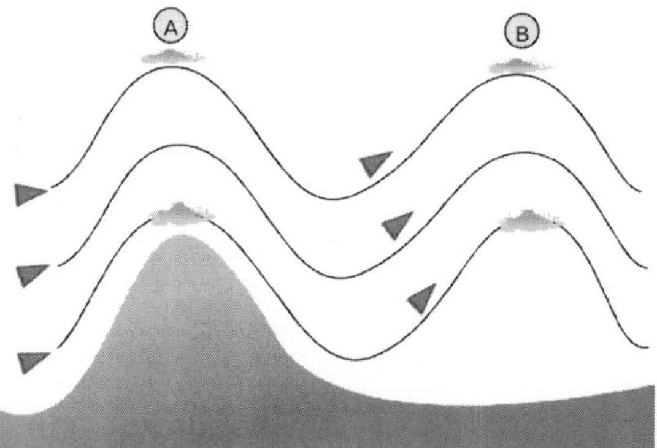

Figure: *The wind flows towards a mountain and produces a first oscillation (A) followed by more waves. The following waves will have lower amplitude because of the natural damping. Lenticular clouds stuck on top of the flow (A) and (B) will appear immobile despite the strong wind.*

Clouds

Both lee waves and the rotor may be indicated by specific wave cloud formations if there is sufficient moisture in the atmosphere, and sufficient vertical displacement to cool the air to the dew point. Waves may also form in dry air without cloud markers. Wave clouds do not move downwind as clouds usually do, but remain fixed in position relative to the obstruction that forms them.

- Around the crest of the wave, adiabatic expansion cooling can form a cloud in shape of a lens (lenticularis). Multiple lenticular clouds can be stacked on top of each other if there are alternating layers of relatively dry and moist air aloft.

- The rotor may generate cumulus or cumulus fractus in its upwelling portion, also known as a "roll cloud". The rotor cloud looks like a line of cumulus. It forms on the lee side and parallel to the ridge line. Its base is near the height of the mountain peak, though the top can extend well above the peak and can merge with the lenticular clouds above. Rotor clouds have ragged leeward edges and are dangerously turbulent.
- A foehn wall cloud may exist at the lee side of the mountains, however this is not a reliable indication of the presence of lee waves.
- A pileus or cap cloud, similar to a lenticular cloud, may form above the mountain or cumulus cloud generating the wave.
- Adiabatic compression heating in the trough of each wave oscillation may also evaporate cumulus or stratus clouds in the airmass, creating a "wave window" or "Foehn gap".

Aviation

Lee waves provide a possibility for gliders to gain altitude or fly long distances when soaring. World record wave flight performances for speed, distance or altitude have been made in the lee of the Sierra Nevada, Alps, Patagonic Andes, and Southern Alps mountain ranges. The Perlan Project is working to demonstrate the viability of climbing above the tropopause in an unpowered glider using lee waves, making the transition into stratospheric standing waves. They did this for the first time on August 30, 2006 in Argentina, climbing to an altitude of 15,460 metres (50,720 ft). The Mountain Wave Project of the Organisation Scientifique et Technique du Vol à Voile focusses on analysis and classification of lee waves and associated rotors.

The conditions favouring strong lee waves suitable for soaring are:

- A gradual increase in windspeed with altitude
- Wind direction within 30° of perpendicular to the mountain ridgeline
- Strong low-altitude winds in a stable atmosphere
- Ridgetop winds of at least 20 knots

The rotor turbulence may be harmful for other small aircraft such as balloons, hang gliders and para gliders. It can even be a hazard for large aircraft; the phenomenon is believed responsible for many aviation accidents and incidents including the in-flight break up of BOAC Flight 911, a Boeing 707, near Mt. Fuji, Japan in 1966, and the in-flight separation of an engine on an Evergreen International Airlines Boeing 747 cargo jet near Anchorage, Alaska in 1993.

The rising air of the wave, which allows gliders to climb to great heights, can also result in high altitude upset in jet aircraft trying to maintain level cruising flight in lee waves. Rising, descending or turbulent air in or above the lee waves can cause overspeed or stall, resulting in mach tuck and loss of control, especially when the aircraft is operated near the "coffin corner".

Other Varieties of Atmospheric Waves

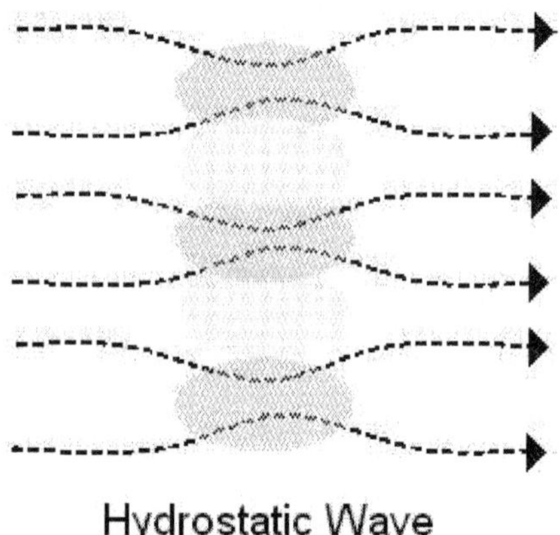

Hydrostatic Wave

Figure: *Hydrostatic wave (schematic drawing)*

There are a variety of distinctive types of waves which form under different atmospheric conditions.

- *Wind shear* can also create waves. This occurs when an atmospheric inversion separates two layers with a marked difference in wind direction. If the wind encounters distortions in the inversion layer caused by thermals coming up from below, it will create significant shear waves in the lee of the distortions that can be used for soaring.
- *Hydraulic jump induced waves* are a type of wave that forms when there exists a lower layer of air which is dense, yet thin relative to the size of the mountain. After flowing over the mountain, a type of shock wave forms at the trough of the flow, and a sharp vertical discontinuity called the hydraulic jump forms which can be several times higher than the mountain. The hydraulic jump is similar to a rotor in that it is very turbulent, yet it is not as spatially localized as a rotor. The hydraulic jump itself acts as an obstruction for the stable layer of air moving above it, thereby triggering wave. Hydraulic jumps can distinguished by their towering roll clouds, and have been observed on the Sierra Nevada range as well as mountain ranges in southern California.

- *Hydrostatic waves* are vertically propagating waves which form over spatially large obstructions. In hydrostatic equilibrium, the pressure of a fluid can depend only on altitude, not on horizontal displacement. Hydrostatic waves get their name from the fact that they approximately obey the laws of hydrostatics, i.e. pressure amplitudes vary primarily in the vertical direction instead of the horizontal. Whereas conventional, non-hydrostatic waves are characterized by horizontal undulations of lift and sink, largely independent of altitude, hydrostatic waves are characterized by undulations of lift and sink at different altitudes over the same ground position.
- *Kelvin–Helmholtz instability* can occur when velocity shear is present within a continuous fluid or when there is sufficient velocity difference across the interface between two fluids.
- *Rossby waves* (or planetary waves) are large-scale motions in the atmosphere whose restoring force is the variation in Coriolis effect with latitude.

Chapter 3

Internal Wave

Internal waves are gravity waves that oscillate within, rather than on the surface of, a fluid medium. To exist, the fluid must be stratified meaning that the density decreases continuously or discontinuously with height due to changes, for example, in temperature and/or salinity. If the density changes over a small vertical distance (as in the case of the thermocline in lakes and oceans or an atmospheric inversion), the waves propagate horizontally like surface waves, but do so at slower speeds as determined by the density difference of the fluid below and above the interface. If the density changes continuously, the waves can propagate vertically as well as horizontally through the fluid.

Internal waves, also called internal gravity waves, go by many other names depending upon the fluid stratification, generation mechanism, amplitude and influence of external forces. If propagating horizontally along an interface where the density rapidly decreases with height, they are specifically called interfacial (internal) waves. If the interfacial waves are large amplitude they are called internal solitary waves or internal solitons. If moving vertically through the atmosphere where substantial changes in air density influences their dynamics, they are called anelastic (internal) waves. If generated by flow over topography, they are called Lee waves or mountain waves. If the mountain waves break aloft, they can result in strong warm winds at the ground known as Chinook winds (in North America) or Foehn winds (in Europe). If generated in the ocean by tidal flow over submarine ridges or the continental shelf, they are called internal tides. If they evolve slowly compared to the Earth's rotational frequency so that their dynamics are influence by the Coriolis effect, they are called inertia gravity waves or, simply, inertial waves. Internal waves are usually distinguished from Rossby waves, which are influenced by the change of Coriolis frequency with latitude.

Visualization of Internal Waves

An internal wave can readily be observed in the kitchen by slowly tilting back and forth a bottle of salad dressing - the waves exist at the interface between oil and vinegar.

Atmospheric internal waves can be visualized by wave clouds: at the wave crests air rises and cools in the relatively lower pressure, which can result in water vapor condensation if the relative humidity is close to 100%. Clouds that reveal internal waves launched by flow over hills are called lenticular clouds because of their lens-like appearance. Less dramatically, a train of internal waves can be visualized by rippled cloud patterns described as herringbone sky or mackerel sky. The outflow of cold air from a thunderstorm can launch large amplitude internal solitary waves at an atmospheric inversion. In northern Australia, these result in Morning Glory clouds, used by some daredevils to glide along like a surfer riding an ocean wave. Satellites over Australia and elsewhere reveal these waves can span many hundreds of kilometres.

Undulations of the oceanic thermocline can be visualized by satellite because the waves increase the surface roughness where the horizontal flow converges and this increases the scatter of sunlight (as in the image at the top of this page showing of waves generated by tidal flow through the Strait of Gibraltar).

Buoyancy, Reduced Gravity and Buoyancy Frequency

According to Archimedes principle, the weight of an immersed object is reduced by the weight of fluid it displaces. This holds for a fluid parcel of density ρ surrounded by an ambient fluid of density ρ_0. Its weight per unit volume is $g(\rho - \rho_0)$, in which g is the acceleration of gravity. Dividing by a characteristic density, ρ_{00}, gives the definition of the reduced gravity:

$$g' \equiv g \frac{\rho - \rho_0}{\rho_{00}}$$

If $\rho > \rho_0$, g' is positive though generally much smaller than g. Because water is much more dense than air, the displacement of water by air from a surface gravity wave feels nearly the full force of gravity ($g' \sim g$). The displacement of the thermocline of a lake, which separates warmer surface from cooler deep water, feels the buoyancy force expressed through the reduced gravity. For example, the density difference between ice water and room temperature water is 0.002 the characteristic density of water. So the reduced gravity is 0.2% that of gravity. It is for this reason that internal waves move in slow-motion relative to surface waves.

Whereas the reduced gravity is the key variable describing buoyancy for interfacial internal waves, a different quantity is used to describe buoyancy in continuously stratified fluid whose density varies with height as $\rho_0(z)$. Suppose a water column is in hydrostatic equilibrium and a small parcel of fluid with density $\rho_0(z_0)$ is displaced vertically by a small distance Δz. The buoyant restoring force results in a vertical acceleration, given by

$$\frac{d^2\Delta z}{dt^2} = -g' = -g(\rho_0(z_0) - \rho_0(z_0 + \Delta z)) / \rho_0(z_0) \simeq -g\left(-\frac{d\rho_0}{dz}\Delta z\right) / \rho_0(z_0)$$

This is the spring equation whose solution predicts oscillatory vertical displacement about in time about with frequency given by the buoyancy frequency:

$$N = \left(-\frac{g}{\rho_0}\frac{d\rho_0}{dz}\right)^{1/2}.$$

The above argument can be generalized to predict the frequency, ω, of a fluid parcel that oscillates along a line at an angle Θ to the vertical:

$$\omega = N\cos\Theta.$$

This is one way to write the dispersion relation for internal waves whose lines of constant phase lie at an angle Θ to the vertical. In particular, this shows that the buoyancy frequency is an upper limit of allowed internal wave frequencies.

Mathematical Modelling of Internal Waves

The theory for internal waves differs in the description of interfacial waves and vertically propagating internal waves. These are treated separately below.

Interfacial Waves

In the simplest case, one considers a two-layer fluid in which a slab of fluid with uniform density ρ_1 overlies a slab of fluid with uniform density ρ_2. Arbitrarily the interface between the two layers is taken to be situated at $z=0$. The fluid in the upper and lower layers are assumed to be irrotational. So the velocity in each layer is given by the gradient of a velocity potential, $\vec{u}=\nabla\phi$, and the potential itself satisfies Laplace's equation:

$$\nabla^2\phi=0$$

Assuming the domain is unbounded and two dimensional (in the x-z plane) and assuming the wave is periodic in x with wavenumber k>0, the equations in each layer reduces to a second-order ordinary differential equation in z. Insisting on bounded solutions the velocity potential in each layer is

$$\phi_1(x,z,t)=Ae^{-kz}\cos(kx-\omega t)$$

and

$$\phi_2(x,z,t)=Ae^{kz}\cos(kx-\omega t)$$

In deriving this structure, matching conditions have been used at the interface requiring continuity of mass and pressure. These conditions also give the dispersion relation:

$$\omega^2=g^{'}k$$

in which the reduced gravity is based up the density difference of the upper and lower layers. Note that the dispersion relation is the same as that for deep water surface waves by setting $g^{'}=g$.

Internal Waves in Uniformly Stratified Fluid

The structure and dispersion relation of internal waves in uniformly stratified, fluid is found through the solution of the linearized conservation of mass, momentum, and internal energy equations assuming the fluid is incompressible and the background density varies by a small amount (the Boussinesq approximation). Assuming the waves are two dimensional in the x-z plane, the respective equations are

$$\partial_x u+\partial_z w$$

$$\rho_{00}\partial_t u=-\partial_x p$$

$$\rho_{00}\partial_t w=-\partial_z p-\rho g$$

$$\partial_t\rho=-w d\rho_0/dz$$

in which ρ is the perturbation density, p is the pressure, and (u,w) is the velocity. The ambient density changes linearly with height as given by $\rho_0(z)$ and ρ_{00}, a constant, is the characteristic ambient density.

Solving the four equations in four unknowns for a wave of the form $\exp[i(kx + mz - \omega t)]$ gives the dispersion relation

$$\omega^2 = N^2 \frac{k^2}{k^2 + m^2} = N^2 \cos^2 \Theta$$

in which N is the buoyancy frequency and $\Theta = \tan^{-1}(m / k)$ is the angle of the wavenumber vector to the horizontal, which is also the angle formed by lines of constant phase to the vertical.

The phase velocity and group velocity found from the dispersion relation predict the unusual property that they are perpendicular and that the vertical components of the phase and group velocities have opposite sign: if a wavepacket moves upward to the right, the crests moves downward to the right.

Internal Waves in the Ocean

Most people think of waves as a surface phenomenon, which acts between water (as in lakes or oceans) and the air. Where low density water overlies high density water in the ocean, internal waves propagate along the boundary. They are especially common over the continental shelf regions of the world oceans and where brackish water overlies salt water at the outlet of large rivers. There is typically little surface expression of the waves, aside from slick bands that can form over the trough of the waves.

Internal waves are the source of a curious phenomenon called dead water, first reported by the Norwegian oceanographer Fridtjof Nansen, in which a boat may experience strong resistance to forward motion in apparently calm conditions. This occurs when the ship is sailing on a layer of relatively fresh water whose depth is comparable to the ship's draft. This causes a wake of internal waves that dissipates a huge amount of energy.

Properties of Internal Waves

Internal waves typically have much lower frequencies and higher amplitudes than surface gravity waves because the density differences (and therefore the restoring forces) within a fluid are usually much smaller. Wavelengths vary from centimetres to kilometres with periods of seconds to hours respectively.

The atmosphere and ocean are continuously stratified: potential density generally increases steadily downward. Internal waves in a continuously stratified medium may propagate vertically as well as horizontally. The dispersion relation for such waves is curious: For a freely-propagating internal wave packet, the direction of propagation of energy (group velocity) is perpendicular to the direction of propagation of wave crests and troughs (phase velocity). An internal wave may also become confined to a finite region of altitude or depth, as a result of varying stratification or wind. Here, the wave is said to be *ducted* or *trapped*, and a vertically standing wave may form, where the vertical component of group velocity approaches zero. A ducted internal wave *mode* may propagate horizontally, with parallel group and phase velocity vectors, analogous to propagation within a waveguide.

At large scales, internal waves are influenced both by the rotation of the Earth as well as by the stratification of the medium. The frequencies of these geophysical wave motions vary from a lower limit of the Coriolis frequency (inertial motions) up to the Brunt–Väisälä frequency, or buoyancy frequency (buoyancy oscillations). Above the Brunt–Väisälä frequency, there may be evanescent internal wave motions, for example those resulting from partial reflection. Internal waves at tidal frequencies are produced by tidal flow over topography/ bathymetry, and are known as internal tides. Similarly, Atmospheric tides arise from, for example, non-uniform solar heating associated with diurnal motion.

Onshore Transport of Planktonic Larvae

Cross-shelf transport, the exchange of water between coastal and offshore environments, is of particular interest for its role in delivering meroplanktonic larvae to often disparate adult populations from shared offshore larval pools. Several mechanisms have been proposed for the cross-shelf of planktonic larvae by internal waves. The prevalence of each type of event depends on a variety of factors including bottom topography, stratification of the water body and tidal influences.

Internal Tidal Bores

Similarly to surface waves, internal waves change as they approach the shore. As the ratio of wave amplitude to water depth becomes such that the wave "feels the bottom," water at the base of the wave slows down due to friction with the sea floor. This causes the wave to become asymmetrical, the face of the wave to steepen and finally the wave will break, propagating forward as an internal bore. Internal waves are often formed as tides pass over a shelf break. The largest of these waves are generated during springtides and those of sufficient magnitude break and progress across the shelf as bores. These bores are evidenced by rapid, step-like changes in temperature and salinity with depth, the abrupt onset of upslope flows near the bottom and packets of high frequency internal waves following the fronts of the bores.

The arrival of cool, formerly deep water associated with internal bores into warm, shallower waters corresponds with drastic increases in phytoplankton and zooplankton concentrations and changes in plankter species abundances. Additionally, while both surface waters and those at depth tend to have relatively low primary productivity, thermoclines are often associated with a chlorophyll maximum layer. These layers in turn attract large aggregations of mobile zooplankton that internal bores subsequently push inshore. Many taxa can be almost absent in warm surface waters, yet plentiful in these internal bores.

Surface Slicks

While internal waves of higher magnitudes will often break after crossing over the shelf break, smaller trains will proceed across the shelf unbroken. At low wind speeds these internal waves are evidenced by the formation of wide surface slicks, oriented parallel to the bottom topography, which progress shoreward with the internal waves. Waters above an internal wave converge and sink in its trough and upwell and diverge over its crest. The convergence zones associated with internal wave troughs often accumulate oils and flotsam that occasionally progress shoreward with the slicks. These rafts of flotsam can also harbor high concentrations of larvae of invertebrates and fish an order of magnitude higher than the surrounding waters.

Predictable Downwellings

Thermoclines are often associated with chlorophyll maximum layers. Internal waves represent oscillations of these thermoclines and therefore have the potential to transfer these phytoplankton rich waters downward, coupling benthic and pelagic systems. Areas affected by these events show higher growth rates of suspension feeding ascidians and bryozoans, likely due to the periodic influx of high phytoplankton concentrations. Periodic depression of the thermocline and associated downwelling may also play an important role in the vertical transport of planktonic larvae.

Trapped Cores

Large steep internal waves containing trapped, reverse-oscillating cores can also transport parcels of water shoreward. These non-linear waves with trapped cores had previously been observed in the laboratory and predicted theoretically. These waves propagate in environments characterized by high shear and turbulence and likely derive their energy from waves of depression interacting with a shoaling bottom further upstream. The conditions favourable to the generation of these waves are also likely to suspend sediment along the bottom as well as plankton and nutrients found along the benthos in deeper water.

Arctic Internal Waves

In the Arctic, surface waves are increasing in both length and amplitude as melting ice leaves areas with greater fetch for the wind to act upon. Internal waves between the surface fresher Arctic water and the deeper saltier Atlantic water can also be expected to be increasing. As internal waves reach shallower water they will break as do surface waves, mixing the layers. Since the deep saltier water is warmer than surface water, this will contribute to more ice melt. Here we have another tipping point.

Atmospheric Gravity Waves

Drop a stone into a pool of water. The spreading ripples are gravity waves. The waves occur between any stable layers of fluids of different density. When the fluid boundary is disturbed, buoyancy forces try to restore the equilibrium. The fluid returns to its original shape, overshoots and oscillations then set in which propagate as waves. Gravity or buoyancy is the restoring force hence the term - gravity waves.

These waves (internal gravity or buoyancy waves) abound in the stable density layering of the upper atmosphere. Their effects are visibly manifest in the curls of the stratosphere's nacreous clouds, in the moving skein-like and billow patterns of the mesosphere's noctilucent clouds and in the slowly shifting bands of the thermosphere's airglow.

What triggers them? The 'stones into the pond' are disturbances far below in the troposphere, for example, wind flow over mountain ranges and violent thunderstorms. Jet stream shear and solar radiation are other sources.

An initial small amplitude at the tropopause increases with height until the waves break in the mesosphere and lower thermosphere. Their wavelengths can range up to thousands of kilometres. Their periods range from a few minutes to days.

They do more than give clouds interesting shapes. They are vital in their role of transferring energy, momentum and chemical species between the different atmospheric layers and in the subsequent influence on upper atmosphere winds, turbulence, temperature and chemistry.

Atmospheric gravity waves are not to be confused with gravitational waves, which are a fluctuation in the curvature of space-time, which propagates as a wave travelling outward from a moving object or system of objects.

Gravity Waves in the Middle and Upper Atmosphere

Gravity is what keeps a planet's gaseous atmosphere from spreading out into space away from the planet.

If we compare the gravitational pull of each planet in our solar system we would find them to be different. This is because a planet's gravity is related to its mass. Usually the greater a planet's mass, the greater the gravitational pull.

As the planet rotates on its axis it exerts a centrifugal effect on the atmosphere. Let's say that you are wearing a pair of lightweight, loose fitting sunglasses. If you spin around quickly in one place, these sunglasses will fly off your nose and move outward away from your face. This is a centrifugal effect.

The air molecules around a planet do the same thing, especially if they are lightweight. Molecules in a gas are in constant motion zipping around and bouncing off each other.

Lightweight gases such as hydrogen and helium move faster than medium-weight gases like nitrogen and oxygen. The heavy gases like carbon dioxide move at a slower rate of speed than the other two.

The gravity on Earth, however, is strong enough to hold onto gases like nitrogen and oxygen. So when the Earth was forming, the lightweight gases of hydrogen and helium escaped into space, leaving behind the heavier gases of nitrogen and oxygen.

Gravitational Potential

In classical mechanics, the gravitational potential at a location is equal to the work (energy transferred) per unit mass that is done by the force of gravity to move an object to a fixed reference location. It is analogous to the electric potential with mass playing the role of charge.

The reference location, where the potential is zero, is by convention infinitely far away from any mass, resulting in a negative potential at any finite distance.

In mathematics the gravitational potential is also known as the Newtonian potential and is fundamental in the study of potential theory.

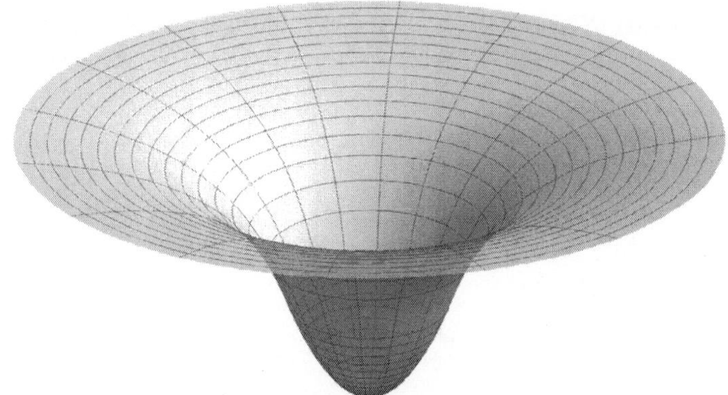

Figure: *Plot of a two-dimensional slice of the gravitational potential in and around a uniform spherical body. The inflection points of the cross-section are at the surface of the body.*

Potential Energy

The gravitational potential (V) is the potential energy (U) per unit mass:

$$U = mV,$$

where m is the mass of the object. Potential energy is equal (in magnitude, but negative) to the work done by the gravitational field moving a body to its given position in space from infinity. If the body has a mass of 1 unit, then the potential energy to be assigned to that body is equal to the gravitational potential. So the potential can be interpreted as the negative of the work done by the gravitational field moving a unit mass in from infinity.

In some situations, the equations can be simplified by assuming a field that is nearly independent of position. For instance, in daily life, in the region close to the surface of the Earth, the gravitational acceleration can be considered constant. In that case, the difference in potential energy from one height to another is to a good approximation linearly related to the difference in height:

$$\Delta U = mg\Delta h.$$

Mathematical Form

The potential V at a distance x from a point mass of mass M can be defined as the work done by the gravitational field bringing a unit mass in from infinity to that point:

$$V(x) = \frac{W}{m} = \frac{1}{m}\int_{\infty}^{x} F\, dx = \frac{1}{m}\int_{\infty}^{x} \frac{GmM}{x^2} dx = -\frac{GM}{x},$$

where G is the gravitational constant. The potential has units of energy per unit mass, e.g., J/kg in the MKS system. By convention, it is always negative where it is defined, and as x tends to infinity, it approaches zero.

The gravitational field, and thus the acceleration of a small body in the space around the massive object, is the negative gradient of the gravitational potential. Thus the negative of a negative gradient yields positive acceleration toward a massive object. Because the potential has no angular components, its gradient is

$$\mathbf{a} = -\frac{GM}{x^3}\mathbf{x} = -\frac{GM}{x^2}\hat{\mathbf{x}},$$

where x is a vector of length x pointing from the point mass toward the small body and $\hat{\mathbf{x}}$ is a unit vector pointing from the point mass toward the small body. The magnitude of the acceleration therefore follows an inverse square law:

$$|\mathbf{a}| = \frac{GM}{x^2}.$$

The potential associated with a mass distribution is the superposition of the potentials of point masses. If the mass distribution is a finite collection of point masses, and if the point masses are located at the points $\mathbf{x}_1, \ldots, \mathbf{x}_n$ and have masses $m_1, \ldots, m_n$, then the potential of the distribution at the point x is

$$V(\mathbf{x}) = \sum_{i=1}^{n} -\frac{Gm_i}{|\mathbf{x} - \mathbf{x}_i|}.$$

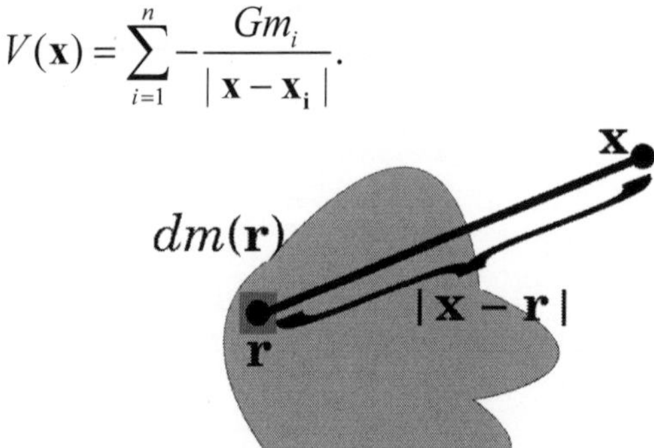

Figure: *Points x and r, with r contained in the distributed mass (gray) and differential mass* dm*(r) located at the point r.*

If the mass distribution is given as a mass measure dm on three-dimensional Euclidean space R^3, then the potential is the convolution of $-G/|r|$ with dm. In good cases this equals the integral

$$V(\mathbf{x}) = -\int_{\mathbf{R}^3} \frac{G}{|\mathbf{x} - \mathbf{r}|} dm(\mathbf{r}),$$

where |x − r| is the distance between the points x and r. If there is a function ρ(r) representing the density of the distribution at r, so that *dm*(r)= ρ(r)*dv*(r), where *dv*(r) is the Euclidean volume element, then the gravitational potential is the volume integral

$$V(\mathbf{x}) = -\int_{\mathbf{R}^3} \frac{G}{|\mathbf{x}-\mathbf{r}|}\rho(\mathbf{r})dv(\mathbf{r}).$$

If V is a potential function coming from a continuous mass distribution ρ(r), then ρ can be recovered using the Laplace operator, Δ:

$$\rho(\mathbf{x}) = \frac{1}{4\pi G}\Delta V(\mathbf{x}).$$

This holds pointwise whenever ρ is continuous and is zero outside of a bounded set. In general, the mass measure *dm* can be recovered in the same way if the Laplace operator is taken in the sense of distributions. As a consequence, the gravitational potential satisfies Poisson's equation.

Spherical Symmetry

A spherically symmetric mass distribution behaves to an observer completely outside the distribution as though all of the mass were concentrated at the centre, and thus effectively as a point mass, by the shell theorem. On the surface of the earth, the acceleration is given by so-called standard gravity *g*, approximately 9.8 m/s^2, although this value varies slightly with latitude and altitude: The magnitude of the acceleration is a little larger at the poles than at the equator because Earth is an oblate spheroid.

Within a spherically symmetric mass distribution, it is possible to solve Poisson's equation in spherical coordinates. Within a uniform spherical body of radius R and density ρ the gravitational force *g* inside the sphere varies linearly with distance r from the centre, giving the gravitational potential inside the sphere, which is

$$V(r) = \frac{2}{3}\pi G\rho(r^2 - 3R^2), \qquad r \le R,$$

which differentiably connects to the potential function for the outside of the sphere.

General Relativity

In general relativity, the gravitational potential is replaced by the metric tensor. When the gravitational field is weak and the sources are moving very slowly compared to light-speed, general relativity reduces to Newtonian gravity, and the metric tensor can be expanded in terms of the gravitational potential.

Gravitational Acceleration § General Relativity

At different points on Earth, objects fall with an acceleration between 9.78 and 9.83 m/s^2 depending on altitude, with a conventional standard value of exactly 9.80665 m/s^2 (approx. 32.174 ft/s^2). Objects with low densities do not accelerate as rapidly due to buoyancy and air resistance.

Newton's Law of Universal Gravitation states that there is a gravitational force between any two masses that is equal in magnitude for each mass, and is aligned to draw the two masses toward each other. The formula is:

$$F = G\frac{m_1 m_2}{r^2}$$

where m_1 and m_2 are the two masses, G is the gravitational constant, and r is the distance between the two masses. The formula was derived for planetary motion where the distances between the planets and the Sun made it reasonable to consider the bodies to be point masses. (For a satellite in orbit, the 'distance' refers to the distance from the mass centres rather than, say, the altitude above a planet's surface.)

If one of the masses is much larger than the other, it is convenient to define a gravitational field around the larger mass as follows:

$$\mathbf{g} = -\frac{GM}{r^2}\hat{\mathbf{r}}$$

where M is the mass of the larger body, and $\hat{\mathbf{r}}$ is a unit vector directed from the large mass to the smaller mass. The negative sign indicates that the force is an attractive force.

In that way, the force acting upon the smaller mass can be calculated as:

$$\mathbf{F} = m\mathbf{g}$$

where $\mathbf{F}$ is the force vector, m is the smaller mass, and g is a vector pointed toward the larger body. Note that has units of acceleration and is a vector function of location relative to the large body, independent of the magnitude (or even the presence) of the smaller mass.

This model represents the "far-field" gravitational acceleration associated with a massive body. When the dimensions of a body are not trivial compared to the distances of interest, the principle of superposition can be used for differential masses for an assumed density distribution throughout the body in order to get a more detailed model of the "near-field" gravitational acceleration. For satellites in orbit, the far-field model is sufficient for rough calculations of altitude versus period, but not for precision estimation of future location after multiple orbits.

The more detailed models include (among other things) the bulging at the equator for the Earth, and irregular mass concentrations (due to meteor impacts) for the Moon. The *Gravity Recovery And Climate Experiment* (GRACE) mission launched in 2002 consists of two probes, nicknamed "Tom" and "Jerry", in polar orbit around the Earth measuring differences in the distance between the two probes in order to more precisely determine the gravitational field around the Earth, and to track changes that occur over time. Similarly, the *Gravity Recovery and Interior Laboratory* (GRAIL) mission from 2011-2012 consisted of two probes ("Ebb" and "Flow") in polar orbit around the Moon to more precisely determine the gravitational field for future navigational purposes, and to infer information about the Moon's physical makeup.

Gravity Model for Earth

The type of gravity model used for the Earth depends upon the degree of fidelity required for a given problem. For many problems such as aircraft simulation, it may be sufficient to consider gravity to be a constant, defined as:

$g = 9.80665$ metres (32.1740 ft) per s^2

based upon data from *World Geodetic System 1984* (WGS-84), where is understood to be pointing 'down' in the local frame of reference.

If it is desirable to model an object's weight on Earth as a function of latitude, one could use the following:

$$g = g_{45} - \tfrac{1}{2}(g_{\text{poles}} - g_{\text{equator}})\cos\left(2lat\frac{\pi}{180}\right)$$

where

- $g_{\text{poles}} = 9.832$ metres (32.26 ft) per s^2
- $g_{45} = 9.806$ metres (32.17 ft) per s^2
- $g_{\text{equator}} = 9.780$ metres (32.09 ft) per s^2
- lat = latitude, between –90 and 90 degrees

Both these models take into account the centrifugal relief that is produced by the rotation of the Earth, and neither accounts for changes in gravity with changes in altitude. It is worth noting that for the mass attraction effect by itself, the gravitational acceleration at the equator is about 0.18% less than that at the poles due to being located farther from the mass centre. When the rotational component is included (as above), the gravity at the equator is about 0.53% less than that at the poles, with gravity at the poles being unaffected by the rotation. So the rotational component change due to latitude (0.35%) is about twice as significant as the mass attraction change due to latitude (0.18%), but both reduce strength of gravity at the equator as compared to gravity at the poles.

Note that for satellites, orbits are decoupled from the rotation of the Earth so the orbital period is not necessarily one day, but also that errors can accumulate over multiple orbits so that accuracy is important. For such problems, the rotation of the Earth would be immaterial unless variations with longitude are modelled. Also, the variation in gravity with altitude becomes important, especially for highly elliptical orbits.

The *Earth Gravitational Model **1996*** (EGM96) contains 130,676 coefficients that refine the model of the Earth's gravitational field (p. 40). The most significant correction term is about two orders of magnitude more significant than the next largest term (p. 40). That coefficient is referred to as the J_2 term, and accounts for the flattening of the poles, or the oblateness, of the Earth. (A shape elongated on its axis-of-symmetry, like an American football, would be called prolate.) A gravitational potential function can be written for the change in potential energy for a unit mass that is brought from infinity into proximity to the Earth. Taking partial derivatives of that function with respect to a coordinate system will then resolve the directional components of the gravitational acceleration vector, as a function of location. The component due to the Earth's rotation can then be included, if appropriate, based on a sidereal day relative to the stars (≈366.24 days/year) rather than on a solar day (≈365.24 days/year). That component is perpendicular to the axis of rotation rather than to the surface of the Earth.

A similar model adjusted for the geometry and gravitational field for Mars can be found in publication NASA SP-8010.

The barycentric gravitational acceleration at a point in space is given by:

$$\mathbf{g} = -\frac{GM}{r^2}\hat{\mathbf{r}}$$

where:

M is the mass of the attracting object, $\hat{\mathbf{r}}$ is the unit vector from centre-of-mass of the attracting object to the centre-of-mass of the object being accelerated, r is the distance between the two objects, and G is the gravitational constant.

When this calculation is done for objects on the surface of the Earth, or aircraft that rotate with the Earth, one has to account that the Earth is rotating and the centrifugal acceleration has to be subtracted from this. For example, the equation above gives the acceleration at 9.820 m/s^2, when GM = 3.986×10^{14} m^3/s^2, and R=6.371×10^6 m. The centripetal radius is $r = R$ cos(*latitude*), and the centripetal time unit is approximately (*day* / 2π), reduces this, for r = 5×10^6 metres, to 9.79379 m/s^2, which is closer to the observed value.

Disregarding air resistance, all small masses dropped simultaneously from the same height will hit the ground at the same time; for example, during Apollo 15 an astronaut on the Moon simultaneously dropped a feather and a hammer and they reached the ground at the same time.

General Relativity

In general relativity the gravitational field is determined by solving the Einstein field equations,

$$\mathbf{G} = \frac{8\pi G}{c^4}\mathbf{T}.$$

Here T is the stress–energy tensor, G is the Einstein tensor, and c is the speed of light,

These equations are dependent on the distribution of matter and energy in a region of space, unlike Newtonian gravity, which is dependent only on the distribution of matter. The fields themselves in general relativity represent the curvature of spacetime. General relativity states that being in a region of curved space is equivalent to accelerating up the gradient of the field. By Newton's second law, this will cause an object to experience a fictitious force if it is held still with respect to the field. This is why a person will feel himself pulled down by the force of gravity while standing still on the Earth's surface. In general the gravitational fields predicted by general relativity differ in their effects only slightly from those predicted by classical mechanics, but there are a number of easily verifiable differences, one of the most well known being the bending of light in such fields.

Multipole Expansion

The potential at a point x is given by

$$V(\mathbf{x}) = -\int_{\mathbb{R}^3} \frac{G}{|\mathbf{x}-\mathbf{r}|}\, dm(\mathbf{r}).$$

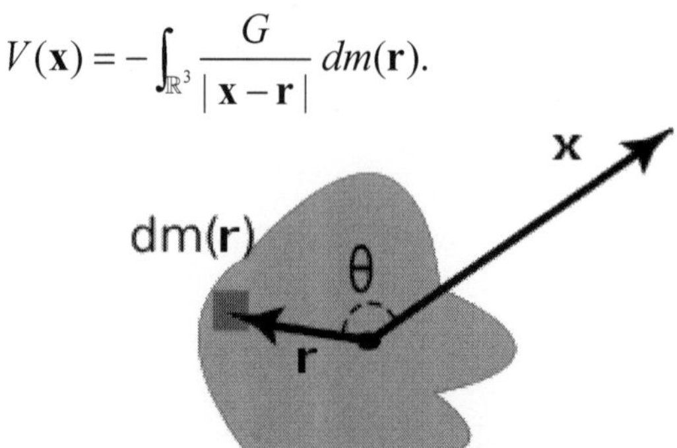

Figure: *Illustration of a mass distribution (grey) with centre of mass as the origin of vectors x and r and the point at which the potential is being computed at the tail of vector x.*

The potential can be expanded in a series of Legendre polynomials. Represent the points x and r as position vectors relative to the centre of mass. The denominator in the integral is expressed as the square root of the square to give

$$\begin{aligned} V(\mathbf{x}) &= -\int_{\mathbb{R}^3} \frac{G}{\sqrt{|\mathbf{x}|^2 - 2\mathbf{x}\cdot\mathbf{r} + |\mathbf{r}|^2}}\, dm(\mathbf{r}) \\ &= -\frac{1}{|\mathbf{x}|}\int_{\mathbb{R}^3} G/\sqrt{1 - 2\frac{r}{|\mathbf{x}|}\cos\theta + \left(\frac{r}{|\mathbf{x}|}\right)^2}\, dm(\mathbf{r}) \end{aligned}$$

where in the last integral, r = | r | and θ is the angle between x and r.

The integrand can be expanded as a Taylor series in $Z = r/|\mathrm{x}|$, by explicit calculation of the coefficients. A less laborious way of achieving the same result is by using the generalized binomial theorem.

The resulting series is the generating function for the Legendre polynomials:

$$\left(1 - 2XZ + Z^2\right)^{-\frac{1}{2}} = \sum_{n=0}^{\infty} Z^n P_n(X)$$

valid for $|X| \leq 1$ and $|Z| < 1$. The coefficients P_n are the Legendre polynomials of degree *n*. Therefore, the Taylor coefficients of the integrand are given by the Legendre polynomials in $X = \cos\theta$. So the potential can be expanded in a series that is convergent for positions x such that $r < |\mathbf{x}|$ for all mass elements of the system (i.e., outside a sphere, centred at the centre of mass, that encloses the system):

$$\begin{aligned} V(\mathbf{x}) &= -\frac{G}{|\mathbf{x}|}\int\sum_{n=0}^{\infty}\left(\frac{r}{|\mathbf{x}|}\right)^n P_n(\cos\theta)dm(\mathbf{r}) \\ &= -\frac{G}{|\mathbf{x}|}\int\left(1+\left(\frac{r}{|\mathbf{x}|}\right)\cos\theta+\left(\frac{r}{|\mathbf{x}|}\right)^2\frac{3\cos^2\theta-1}{2}+\cdots\right)dm(\mathbf{r}) \end{aligned}$$

The integral $\int r\cos\theta dm$ is the component of the centre of mass in the x direction; this vanishes because the vector x emanates from the centre of mass. So, bringing the integral under the sign of the summation gives

$$V(\mathbf{x}) = -\frac{GM}{|\mathbf{x}|}-\frac{G}{|\mathbf{x}|}\int\left(\frac{r}{|\mathbf{x}|}\right)^2\frac{3\cos^2\theta-1}{2}dm(\mathbf{r})+\cdots$$

This shows that elongation of the body causes a lower potential in the direction of elongation, and a higher potential in perpendicular directions, compared to the potential due to a spherical mass, if we compare cases with the same distance to the centre of mass. (If we compare cases with the same distance to the *surface* the opposite is true.)

Numerical Values

The absolute value of gravitational potential at a number of locations with regards to the gravitation from the Earth, the Sun, and the Milky Way is given in the following table; i.e. an object at Earth's surface would need 60 MJ/kg to "leave" Earth's gravity field, another 900 MJ/kg to also leave the Sun's gravity field and more than 130 GJ/kg to leave the gravity field of the Milky Way. The potential is half the square of the escape velocity.

Location	*W.r.t. Earth*	*W.r.t. Sun*	*W.r.t. Milky Way*
Earth's surface	60 MJ/kg	900 MJ/kg	≥ 130 GJ/kg
LEO	57 MJ/kg	900 MJ/kg	≥ 130 GJ/kg
Voyager 1 (17,000 million km from Earth)	23 J/kg	8 MJ/kg	≥ 130 GJ/kg
0.1 light-year from Earth	0.4 J/kg	140 kJ/kg	≥ 130 GJ/kg

Compare the gravity at these locations.

The Gravity Method

For human exploration of the solar system, instruments must meet criteria of low mass, low volume, low power demand, safe operation, and ruggedness and reliability. Tools used for planetary exploration will need to address fundamental scientific questions and identify precious resources, such as water.

The primary goal of studying detailed gravity data is to provide a better understanding of the subsurface geology. The gravity method is a relatively cheap, non-invasive, non-destructive remote sensing method that has already been tested on the lunar surface. It is also passive – that is, no energy need be put into the ground in order to acquire data; thus, the method is well suited to a populated setting such as Taos, and a remote setting such as Mars. The small portable instrument also permits walking traverses – ideal, in view of the congested tourist traffic in Taos.

Measurements of gravity provide information about densities of rocks underground. There is a wide range in density among rock types, and therefore geologists can make inferences about the distribution of strata. In the Taos Valley, we are attempting to map subsurface faults. Because faults commonly juxtapose rocks of differing densities, the gravity method is an excellent exploration choice.

Wave Stress Parameterization

Parametrization (or parameterization; also parameterisation, parametrisation) is the process of deciding and defining the parameters necessary for a complete or relevant specification of a model or geometric object.

Parametrization is also the process of finding parametric equations of a curve, a surface, or, more generally, a manifold or a variety, defined by an implicit equation. The inverse process is called implicitization.

Sometimes, this may only involve identifying certain parameters or variables. If, for example, the model is of a wind turbine with a particular interest in the efficiency of power generation, then the parameters of interest will probably include the number, length and pitch of the blades.

Most often, parametrization is a mathematical process involving the identification of a complete set of effective coordinates or degrees of freedom of the system, process or model, without regard to their utility in some design. Parametrization of a line, surface or volume, for example, implies identification of a set of coordinates that allows one to uniquely identify any point (on the line, surface, or volume) with an ordered list of numbers. Each of the coordinates can be defined parametrically in the form of a parametric curve (one-dimensional) or a parametric equation (2+ dimensions).

Non-Uniqueness

Parametrizations are not generally unique. The ordinary three-dimensional object can be parametrized (or 'coordinatized') equally efficiently with Cartesian coordinates (x,y,z), cylindrical polar coordinates (ρ, φ, z), spherical coordinates (r,φ,θ) or other coordinate systems.

Similarly, the colour space of human trichromatic colour vision can be parametrized in terms of the three colours red, green and blue, RGB, or with cyan, magenta, yellow and black, CMYK.

Dimensionality

Generally, the minimum number of parameters required to describe a model or geometric object is equal to its dimension, and the scope of the parameters—within their allowed ranges—is the parameter space. Though a good set of parameters permits identification of every point in the parameter space, it may be that, for a given parametrization, different parameter values can refer to the same 'physical' point. Such mappings are surjective but not injective. An example is the pair of cylindrical polar coordinates (ρ,φ,z) and $(\rho,\varphi + 2\pi,z)$.

Parametrization Invariance

As indicated above, there is arbitrariness in the choice of parameters of a given model, geometric object, etc. Often, one wishes to determine intrinsic properties of an object that do not depend on this arbitrariness, which are therefore independent of any particular choice of parameters. This is particularly the case in physics, wherein parametrization invariance (or 'reparametrization invariance') is a guiding principle in the search for physically acceptable theories (particularly in general relativity).

For example, whilst the location of a fixed point on some curved line may be given by a set of numbers whose values depend on how the curve is parametrized, the length (appropriately defined) of the curve between *two* such fixed points will be independent of the particular choice of parametrization (in this case: the method by which an arbitrary point on the line is uniquely indexed). The length of the curve is a parameterization-invariant quantity therefore. In such cases parameterization is a mathematical tool employed to extract a result whose value does not depend on, or make reference to, the details of the parameterization. More generally, parametrization invariance of a physical theory implies that either the dimensionality or the volume of the parameter space is larger than is necessary to describe the physics (the quantities of physical significance) in question.

Though the theory of General Relativity can be expressed without reference to a coordinate system, calculations of physical (i.e. observable) quantities such as the curvature of spacetime invariably involve the introduction of a particular coordinate system in order to refer to spacetime points involved in the calculation. In the context of General Relativity then, the choice of coordinate system may be regarded as a method of 'parameterizing' the spacetime, and the insensitivity of the result of a calculation of a physically-significant quantity to that choice can be regarded as an example of parameterization invariance.

As another example, physical theories whose observable quantities depend only on the *relative* distances (the ratio of distances) between pairs of objects are said to be scale invariant. In such theories any reference in the course of a calculation to an absolute distance would imply the introduction of a parameter to which the theory is invariant.

It is now well established that terrain of even gentle relief can generate gravity waves in the night-time atmosphere. These waves propagate upward, and under very common conditions they will grow in amplitude until becoming convectively unstable. Because linear wave theory does not encompass wave breakdowns, this process must be parameterized. This parameterization has been used to explain the observations of turbulence in the residual layer where turbulence is not expected. However, to date this wave-stress parameterization has been confined to kinematic models, and so it is not clear how the wave-stress divergence impacts time-dependent models. This question is explored using a single-column version of a mesoscale model.

Density of Earth Materials

Earth materials is a general term that includes minerals, rocks, soil and water. These are the naturally occurring materials found on Earth that constitute the raw materials upon which our global society exists. Earth materials are vital resources that provide the basic components for life, agriculture and industry.

Earth Materials can also Include Metals and Precious Rocks

The type of materials available locally will of course vary depending upon the conditions in the area of the building site.

In many areas, indigenous stone is available from the local region, such as limestone, marble, granite, and sandstone. It mat be cut in quarries or removed from the surface of the ground (flag and fieldstone). Ideally, stone from the building site can be utilized. Depending on the stone type, it can be used for structural block, facing block, pavers, and crushed stone.

Most brick plants are located near the clay source they use to make brick. Bricks are molded and baked blocks of clay. Brick products come in many forms, including structural brick, face brick, roof tile, structural tile, paving brick, and floor tile.

Caliche is a soft limestone material which is mined from areas with calcium-carbonate soils and limestone bedrock. It is best known as a road bed material, but it can be processed into an unfired building block, stabilized with an additive such as cement. Other earth materials include soil blocks typically stabilized with a cement additive and produced with forms or compression.

Rammed Earth consists of walls made from moist, sandy soil, or stabilized soil, which is tamped into form work. Walls are a minimum of 123 thick. Soils should contain about 30% clay and 70% sand.

Considerations

The use of locally available and indigenous earth materials has several advantages in terms of sustainability. They are:

Reduction of energy costs related to transportation. Reduction of material costs due to reduced transportation costs, especially for well-established industries. Support of local businesses and resource bases. Care must be taken to ensure that non-renewable earth materials are not over-extracted. Ecological balance within the region needs to be maintained while efficiently utilizing its resources. Many local suppliers carry materials that have been shipped in from out of the area, so it is important to ask for locally produced/quarried materials.

Both brick and stone materials are aesthetically pleasing, durable, and low maintenance. Exterior walls weather well, eliminating the need for constant refinishing and sealing. Interior use of brick and stone can also provide excellent thermal mass, or be used to provide radiant heat. Some stone and brick makes an ideal flooring or exterior paving material, cool in summer and possessing good thermal properties for passive solar heating. Caliche block has been produced for applications similar to stone and brick mentioned above. Caliche or earth material block has special structural and finishing characteristics.

Rammed earth is more often considered for use in walls, although it can also be used for floors. Rammed earth and caliche block can be used for structural walls, and offer great potential as low-cost material alternatives with low embodied energy. In addition, such materials are fireproof.

Caliche block and rammed earth can be produced on-site. It is very important to have soils tested for construction material use. Some soils, such as highly expansive or bentonite soils, are not suitable for structural use. Testing labs are available in most areas to determine material suitability for structural use and meeting codes.

Soils for traditional adobe construction are not found in some areas, but other soils for earth building options are available. Many areas have a high percentage of soils suitable for ramming. Caliche is also abundant in many areas (covering 14% of the Austin geographic area, for instance) and is readily available locally.

Gravity Data Acquisition

Gravity is a potential field, i.e., it is a force that acts at a distance. The gravity method is a non-destructive geophysical technique that measures differences in the earth's gravitational field at specific locations. It has found numerous applications in engineering, environmental and geothermal studies including locating voids, faults, buried stream valleys, water table levels and geothermal heat sources. The success of the gravity method depends on the different earth materials having different bulk densities (mass) that produce variations in the measured gravitational field. These variations can then be interpreted by a variety of analytical and computers methods to determine the depth, geometry and density that causes the gravity field variations. For better definition of the bodies causing the perturbations in the gravity field, the gravity data should be collected with small stations spacing, such as 1 km. For engineering investigations this may be as low as 5 metres or less. In addition, gravity station elevations must be determined to within 0.2 metres. Using the highly precise locations and elevations plus all other quantifiable disturbing effects, the data are processed to remove all these predictable effects. The most commonly used processed data are known as Bouguer gravity anomalies, measured in mGal. The interpretation of Bouguer gravity anomalies ranges from just manually inspecting the grid or profiles for variations in the gravitational field to more complex methods that involves separating the gravity anomaly due to an object of interest from some sort of regional gravity field. From this, bodies and structures can be inferred which may be of geothermal interest.

Volcanic centres, where geothermal activity is found, are indicators of cooling magma or hot rock beneath these areas as shown by the recent volcanic flows, ashes, volcanic domes and abundant hydrothermal activities in the form fumaroles and hot springs. Gravity studies in volcanic areas have effectively demonstrated that this method provides good evidence of shallow subsurface density variations, associated with the structural and magmatic history of a volcano. There is a correlation between gravity highs with centres of recent volcanism, intensive faulting and geothermal activity. For example, in the Kenya rift, Olkaria, Domes and Suswa geothermal centres are located on the crest of a gravity high.

Instrumentation

A gravity metre or gravimetre measures the variations in the earth's gravitational field. The variations in gravity are due to lateral changes in the density of the subsurface rocks in the vicinity of the measuring point. Because the density variations are very small and uniform, the gravimetres have to be very sensitive so as to measure one part in 100 million of the earth's gravity field (980 gals or 980,000 mgals) in units of mgals or microgals.

The most commonly used metres do not measure an absolute gravitational acceleration but differences in relative acceleration. There are several gravity metre manufacturers where the accuracies of these metres can vary greatly. The common gravimetres on the market are the Worden gravimetre, the Scintrex and the La Coste Romberg gravimetre. Since these metres are temperature controlled and contain small pen lights to read the metres, they are connected to rechargeable batteries. The metre usually has two batteries, which allows for over 16 hours of readings. The Worden gravimetre is an entirely mechanical and optical relying only on an AA battery for illuminating the crosshairs.

It uses a fixed length spring and mass attached to a calibration spring and veneer scale to measure gravitational acceleration. The Scintrex Autograv is semi-automated, but although a bit more expensive, it has been shown to have a higher stability and experience less tares (a sudden jump in a gravity reading) over long periods of time. The Lacoste and Romberg model gravity metre has an advertised repeatability of 3 microgals (980,000,000 microgals is the Earth's gravitational field) and is one of the preferred instruments for conducting gravity surveys in industry.

Data Acquisition

Gravity data acquisition is a relatively simple task that can be performed by one person. However, two people are usually necessary to determine the location (latitude, longitude and elevation) of the gravity stations. In order to detect a target (e.g., a dense body or fault), gravity readings must be taken along traverses that cross the location of the target. Its expected size will determine the distance between readings (station spacing), with larger station separations for large target and small separations for small ones. It is usually advisable to model the expected anomaly mathematically before conducting

Field work: From this, along with the expected instrument accuracy, an estimate can be made of the anomaly size and the required station spacing.

Surveys are conducted by taking gravity readings at regular intervals along a traverse that crosses the expected location of the target. However, in order to take into account the expected drift of the instrument, one station (a local base station) must be located and has to be reoccupied every half to 1 hour or so (depending on the instrument drift characteristics) to obtain the natural drift of the instrument. These repeated readings are performed because even the most stable gravity metre will have their readings drift with time due to elastic creep within the metre's springs and also to help remove the gravitational effects of the earth tides readings have to be taken at a base station. The instrument drift is usually linear and less than 0.01 mGal/hour under normal operating conditions. Since gravity decreases as elevation increases, the elevation of each station has to be measured with an error of no more than about 3 cm. Readings are taken by placing the instrument on the ground and levelling it. This may be automatic with some instruments, as it is with the Scintrex.

An important factor in obtaining useful gravity values in detailed surveys is determining the earth tide effect as their gravitational effects may be greater than the gravity field variations due to the anomalous features being sought. The final aspect of reading a gravity metre concerns seismic activity or cultural movement such as those of vehicles or people. These will disrupt the readings (the metre is actually a low-frequency seismometre) and even though the Scintrex metre has an anti-seismic filter.

Romberg metres are also mechanically damped to lessen the effects of earthquakes), readings will still be disrupted. If this occurs, all operations should be stopped. Experience indicates that one should wait at least 1-2 hours after a seismic event before resuming the survey. From the drift curve, a base reading corresponding to the time a particular gravity station was measured is obtained by subtracting the base reading from the station reading. This gravity reading is not in mGal but in gravity metre units. One must multiply the gravity metre reading by the manufacturer supplied metre constants (calibration constants) to obtain mGal.

In addition to obtaining a gravity reading, a horizontal position and the elevation of the gravity station must be obtained. The horizontal position could be either latitude and longitude or the x and y distances (metres or feet) from a predetermined origin. The required elevation accuracy for detailed surveys is between 0.004 and 0.2 m and to obtain such accuracy requires performing either an electronic distance metre (theodolite) survey or a total-field differentially corrected global positioning survey (GPS).

Gravity Data Processing

The last task of most fieldwork is to determine the topographic changes and the effects of buildings surrounding a gravity station. Both of these effects will be used later in processing the gravity data.

There are a number of techniques to determine the elevation changes and these usually involve a combination of recording elevation changes in the field and computer computations using digital elevation models (DEM). The most common technique is by Hammer where one records an elevation change in quadrants at set distances (commonly from 0 to 1000 metres) from the gravity station. A newly developed technique uses a laser-positioning gun to obtain more accurate elevation changes within 100 metres of a gravity station. The best technique is to use Hammer's method for near (up to 200 metres) station elevation changes and computer methods based on accurate (at least 10 metre 7.5 minute quadrangles) Dam's.

The observed gravity readings obtained from the gravity survey reflect the gravitational field due to all masses in the earth and the effect of the earth's rotation. Several corrections have to be applied to the field gravity readings. To interpret gravity data, one must remove all known gravitational effects not related to the subsurface density changes. Each reading has to be corrected for elevation, the influence of tides, latitude and, if significant local topography exists, a topographic correction. To understand the corrections, a gravity reading is first considered on the surface of a flat ground surface. Corrections are then applied to account for deviations from this condition. As the distance from the centre of the earth increases, the pull of gravity decreases. The correction to account for this is called the elevation correction. If the gravity reading is taken on top of a hill, then there is a deficit of mass on either side of the hill, compared to a horizontal ground surface. This is corrected with a topographic correction. It is important to realise that mass higher than the reading site will also influence the data and has to be accounted for. This may occur in areas of significant topography or when surveys are conducted near large buildings. In a modern instrument such as the Scintrex, the metre may automatically apply the tidal and drift corrections.

The difference between observed gravity (gobs) and theoretical gravity (gth) at any point on the Earth's surface after reducing the gravity readings to the geoidal surface (i.e., making all the required corrections) is known as the Bouguer gravity anomaly or Bouguer gravity and the results are now due to lateral variations in density in the subsurface (assumed to be caused by geologic structure being sought) which can be used for interpretations.

Data Analysis and Interpretation

The object of the gravity method is to determine information about the earth's subsurface. One can just carry out a qualitative examination of the grid of gravity values, contour maps or the gravity profiles to determine the lateral location of any gravity variations or one can perform a more detailed analysis in order to quantify the nature (depth, geometry, density) of the subsurface feature causing the gravity variations. To determine the later, it is usually necessary to separate the anomaly of interest (residual) from the remaining background anomaly (regional). Then the residual gravity anomaly is modelled to determine the depth, density and geometry of the anomaly's source.

Most other regional-residual anomaly separation techniques involve mathematical operations using a computer. One problem with the mathematical techniques is that they do not accurately represent the "true" residual gravity anomaly due to a specific body. Thus, they should not be used for quantitative interpretation of the subsurface but only for qualitative interpretation. The most common mathematical techniques are surface fitting and weighted averaging. Once the residual has been removed from the Bouguer gravity data modelling can be done over the feature of interest.

Gravity modelling is usually the final step in gravity interpretation and involves trying to determine the density, depth and geometry of one or more subsurface bodies. The modelling procedure commonly involves using a residual gravity anomaly. When modelling a residual gravity anomaly, the interpreter must use a density contrast between the body of interest and the surrounding material, while modelling Bouguer gravity anomalies; the density of the body is used.

There are many different techniques available to perform the modelling procedure and they can be broken down into three main categories:

1) analytical solutions due to simple geometries,
2) forward modelling using 2 -(two-dimensional), 2.5-(two and one-half dimensional) and 3-D (three-dimensional) irregularly shaped bodies, and
3) inverse modelling using 2-, 2.5-and 3-D irregularly shaped bodies.

Most of these techniques involve iterative modelling (by the aid of a computer), where the gravitational field due to the model is calculated and compared to the observed or residual gravity anomalies. If the calculated values do not match the observed anomalies, the model is changed and the procedure is performed again until the match between the calculated values and the observed anomalies is deemed close enough.

Micro-Gravity Monitoring

Reservoir engineering calculations of mass and energy balance on producing geothermal reservoirs require information about in-and out-flows from the reservoir. Such information is usually available for surface flows, such as production, injection, and natural discharge. Values for subsurface in-or out-flows are difficult to get. One method used is a history matching process whereby reservoir performance is computed for various strengths of influx and the matched against observed performance.

A more direct and independent method is through repeat micro-gravity over a producing field. As mass is removed from a geothermal reservoir the gravity field above the reservoir will change. For an influx it will increase while for a loss it will decrease. By measuring the surface gravity field at two points in time the change in gravity over the reservoir during the time interval can be determined.

Maximum gravity changes show a constant trend in time, but different characteristic distributions from zone to zone. This information has been correlated with production data (enthalpy and mass output) from nearby wells as well as assisting in identifying zones for re-injection.

Gravity Anomaly Interpretation

A gravity anomaly is the difference between the observed acceleration of a planet's gravity and a value predicted from a model. A location with a positive anomaly exhibits more gravity than predicted, while a negative anomaly exhibits a lower value than predicted. The anomaly is the body or effect that causes the deviation from the "ideal" gravity model. Many data corrections must be made to the measured gravity value in order to extract the response of the local anomaly, or local geology, which is typically the goal of applied geophysics.

Causes

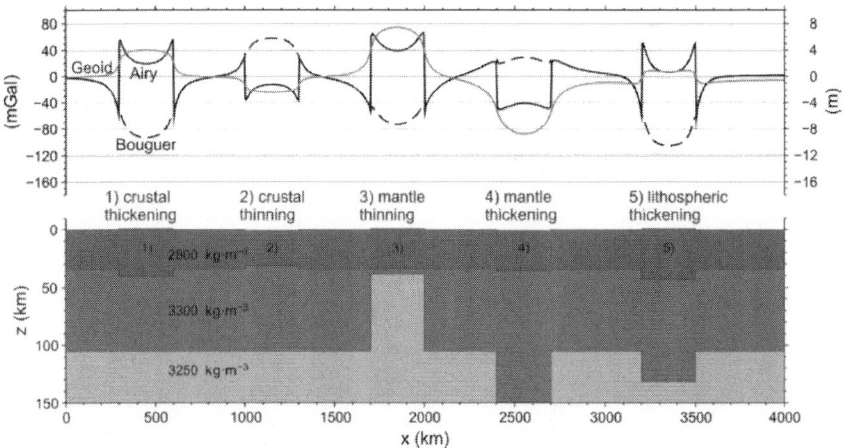

Figure: *Gravity and geoid anomalies caused by various crustal and lithospheric thickness changes relative to a reference configuration. All settings are under local isostatic compensation.*

Lateral variations in gravity anomalies are related to anomalous density distributions within the Earth. Gravity measures help us to understand the internal structure of the planet. Synthetic calculations show that the gravity anomaly signature of a thickened crust (for example, in orogenic belts produced by continental collision) is negative and larger in absolute value, relative to a case where thickening affects the entire lithosphere.

The Bouguer anomalies usually are negative in the mountains because of isostasy: the rock density of their roots is lower, compared with the surrounding earth's mantle. Typical anomalies in the Central Alps are –150 milligals (–1.5 mm/s^2).

Rather local anomalies are used in applied geophysics: if they are positive, this may indicate metallic ores. At scales between entire mountain ranges and ore bodies, Bouguer anomalies may indicate rock types. For example, the northeast-southwest trending high across central New Jersey represents a graben of Triassic age largely filled with dense basalts.

Salt domes are typically expressed in gravity maps as lows, because salt has a low density compared to the rocks the dome intrudes. Anomalies can help to distinguish sedimentary basins whose fill differs in density from that of the surrounding region.

Geodesy and Geophysics

In geodesy and geophysics, the usual theoretical model is the gravity on the surface of a reference ellipsoid such as WGS84.

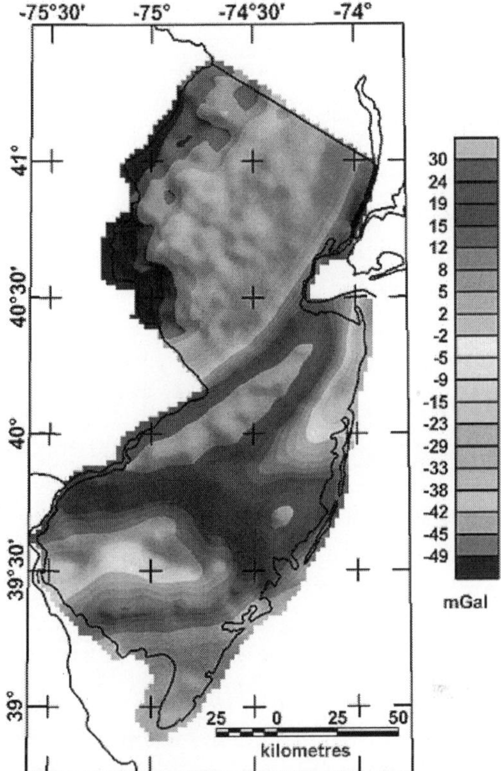

Figure: *(Bouguer) gravity anomaly map of the state of New Jersey (USGS)*

To understand the nature of the gravity anomaly due to the subsurface, a number of corrections must be made to the measured gravity value:

1. The *theoretical gravity* (smoothed normal gravity) should be removed in order to leave only local effects.
2. The elevation of the point where each gravity measurement was taken must be reduced to a reference datum to compare the whole profile. This is called the *Free-air Correction*, and when combined with the removal of theoretical gravity leaves the free-air anomaly.
3. the normal gradient of gravity (rate of change of gravity by change of elevation), as in free air, usually 0.3086 milligals per metre, or the *Bouguer gradient* of 0.1967 mGal/m (19.67 μm/

($s^2 \cdot m$) which considers the mean rock density (2.67 g/cm^3) beneath the point; this value is found by subtracting the gravity due to the Bouguer plate, which is 0.1119 mGal/m (11.19 μm/($s^2 \cdot m$)) for this density. Simply, we have to correct for the effects of any material between the point where gravimetry was done and the geoid. To do this we model the material in between as being made up of an infinite number of slabs of thickness t. These slabs have no lateral variation in density, but each slab may have a different density than the one above or below it. This is called the *Bouguer correction.*

4. and (in special cases) a digital terrain model (DTM). A terrain correction, computed from a model structure, accounts for the effects of rapid lateral change in density, e.g. edge of plateau, cliffs, steep mountains, etc.

For these reductions, different methods are used:

- The gravity changes as we move away from the surface of the Earth. For this reason, we must compensate with the free-air anomaly (or Faye's anomaly): application of the *normal gradient* 0.3086 mGal/m, but no terrain model. This anomaly means a downward shift of the point, together with the whole shape of the terrain. This simple method is ideal for many geodetic applications.
- simple Bouguer anomaly: downward reduction just by the Bouguer gradient (0.1967). This anomaly handles the point as if it is located on a flat plain.
- refined (or complete) Bouguer anomaly (usual abbreviation Δg_B): the DTM is considered as accurate as possible, using a standard density of 2.67 g/cm^3 (granite, limestone). Bouguer anomalies are ideal for geophysics because they show the effects of different rock densities in the subsurface.
    - The difference between the two - the differential gravitational effect of the unevenness of the terrain - is called the terrain effect. It is always negative (up to 100 milligals).
    - The difference between *Faye anomaly* and Δg_B is called *Bouguer reduction* (attraction of the terrain).
- special methods like that of Poincare-Prey, using an interior gravity gradient of about 0.009 milligal per metre (90 nm/($s^2 \cdot m$)). These methods are valid for the gravity within boreholes or for special geoid computations.

Satellite Measurements

Large-scale gravity anomalies can be detected from space. The Gravity Recovery and Climate Experiment (GRACE) consists of two satellites that can detect gravitational changes across the Earth.

Gravity Recovery and Climate Experiment

The Gravity Recovery And Climate Experiment (GRACE), a joint mission of NASA and the German Aerospace Centre, has been making detailed measurements of Earth's gravity field anomalies since its launch in March 2002.

Gravity is determined by mass. By measuring gravity anomalies, GRACE shows how mass is distributed around the planet and how it varies over time. Data from the GRACE satellites is an important tool for studying Earth's ocean, geology, and climate.

GRACE is a collaborative endeavour involving the Centre for Space Research at the University of Texas, Austin; NASA's Jet Propulsion Laboratory, Pasadena, Calif.; the German Space Agency and Germany's National Research Centre for Geosciences, Potsdam. The Jet Propulsion Laboratory is responsible for the overall mission management under the NASA ESSP program.

The principal investigator is Dr. Byron Tapley of the University of Texas Centre for Space Research, and the co-principal investigator is Dr. Christoph Reigber of the GeoForschungsZentrum (GFZ) Potsdam.

The GRACE satellites were launched from Plesetsk Cosmodrome, Russia on a Rockot (SS-19 + Breeze upper stage) launch vehicle, on March 17, 2002.

As of November 2012 the craft was expected to remain in a slowly decaying orbit until 2015 or 2016.

Discoveries and Applications

The monthly gravity anomalies maps generated by Grace are up to 1,000 times more accurate than previous maps, substantially improving the accuracy of many techniques used by oceanographers, hydrologists, glaciologists, geologists and other scientists to study phenomena that influence climate.

From the thinning of ice sheets to the flow of water through aquifers and the slow currents of magma inside Earth, measurements of the amount of mass involved provided by GRACE help scientists better understand these important natural processes. Among the first important applications for GRACE data was to improve the understanding of global ocean circulation. The hills and valleys in the ocean's surface are due to currents and variations in Earth's gravity field. GRACE enables separation of those two effects to better measure ocean currents and their effect on climate. GRACE data are also critical in helping to determine the cause of sea level rise, whether it is the result of mass being added to the ocean, from melting glaciers, for example, or from thermal expansion of warming water or changes in salinity.

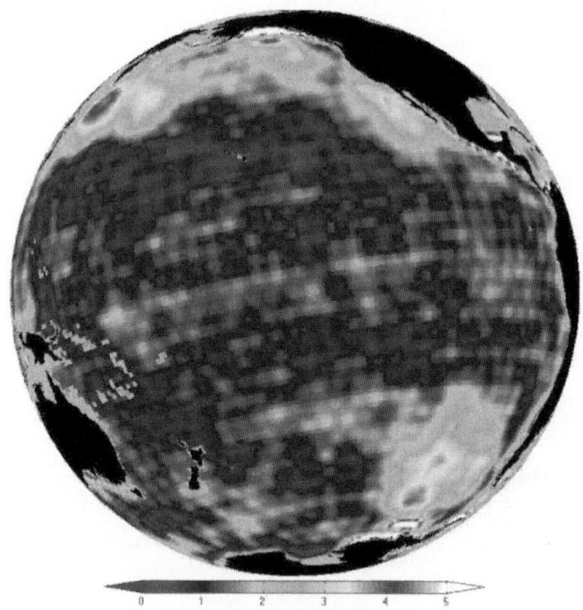

Figure: *Variations in ocean bottom pressure measured by GRACE*

As of February 2012, the data obtained by GRACE are the most precise gravimetric data yet recorded: they have been used to re-analyse data obtained from the LAGEOS experiment to try to measure the relativistic frame-dragging effect.

In 2006, a team of researchers led by Ralph von Frese and Laramie Potts used GRACE data to discover the 480-kilometre (300 mi) wide Wilkes Land crater in Antarctica, which probably formed about 250 million years ago. GRACE has been used to map the hydrologic cycle in the Amazon River basin and the location and magnitude of post-glacial rebound from changes in the free air gravity anomaly. GRACE data has also been used to analyze the shifts in the Earth's crust caused by the earthquake that created the 2004 Indian Ocean tsunami. Scientists have recently developed a new way to calculate ocean bottom pressure—as important to oceanographers as atmospheric pressure is to meteorologists—using GRACE data.

How GRACE Works

GRACE is the first Earth-monitoring mission in the history of space flight whose key measurement is not derived from electromagnetic waves either reflected off, emitted by, or transmitted through Earth's surface and/or atmosphere. Instead, the mission uses a microwave ranging system to accurately measure changes in the speed and distance between two identical spacecraft flying in a polar orbit about 220 kilometres (140 mi) apart, 500 kilometres (310 mi) above Earth. The ranging system is sensitive enough to detect separation changes as small as 10 micrometres (approximately one-tenth the width of a human hair) over a distance of 220 kilometres.

As the twin GRACE satellites circle the globe 15 times a day, they sense minute variations in Earth's gravitational pull. When the first satellite passes over a region of slightly stronger gravity, a gravity anomaly, it is pulled slightly ahead of the trailing satellite. This causes the distance between the satellites to increase. The first spacecraft then passes the anomaly, and slows down again; meanwhile the following spacecraft accelerates, then decelerates over the same point.

By measuring the constantly changing distance between the two satellites and combining that data with precise positioning measurements from Global Positioning System (GPS) instruments, scientists can construct a detailed map of Earth's gravity anomalies.

The two satellites (nicknamed "Tom" and "Jerry") constantly maintain a two-way microwave-ranging link between them. Fine distance measurements are made by comparing frequency shifts of the link. As a cross-check, the vehicles measure their own movements using accelerometres. To establish baseline positions and fulfill housekeeping functions, the satellites also use star cameras, magnetometres, and GPS receivers. The GRACE vehicles also have optical corner reflectors to enable laser ranging from ground stations, bridging the range between spacecraft positions and Doppler ranges.

Spacecraft

The spacecraft were manufactured by Astrium of Germany, using its "Flexbus" platform. The microwave RF systems, and attitude determination and control system algorithms were provided by Space Systems/Loral. The star cameras used to measure the spacecraft attitude were provided by Technical University of Denmark. The instrument computer along with a highly precise BlackJack GPS receiver and digital signal processing system has been provided by JPL in Pasadena. The highly precise accelerometre that is needed to separate atmospheric and solar wind effects from the gravitation data was manufactured by ONERA.

GRACE Follow On

The GeoForschungsZentrum (GFZ) Potsdam has announced a follow on of the GRACE mission. GRACE-FO mission will be a collaboration between GFZ and NASA and is scheduled to be launched in August 2017 on a Dnepr from Baikonur Cosmodrome. The orbit and the design of GRACE-FO will be very similar to GRACE; the distance between the two spacecraft of GRACE-FO will be measured also with lasers (the original GRACE used microwave ranging) as a technological experiment in preparation for future satellites.

Geoid

The geoid is the shape that the surface of the oceans would take under the influence of Earth's gravitation and rotation alone, in the absence of other influences such as winds and tides. This surface is extended through the continents (such as with very narrow hypothetical canals). All points on the geoid have the same gravitational potential. The force of gravity acts everywhere perpendicular to the geoid, meaning that plumb lines point perpendicular and water levels parallel to the geoid.

Specifically, the geoid is the equipotential surface that would coincide with the mean ocean surface of the Earth if the oceans and atmosphere were in equilibrium, at rest relative to the rotating Earth, and extended through the continents (such as with very narrow canals). According to Gauss, who first described it, it is the "mathematical figure of the Earth", a smooth but highly irregular surface whose shape results from the uneven distribution of mass within and on the surface of the Earth. It does not correspond to the actual surface of the Earth's crust, but to a surface which can only be known through extensive gravitational measurements and calculations. Despite being an important concept for almost two hundred years in the history of geodesy and geophysics, it has only been defined to high precision in recent decades. It is often described as the true physical figure of the Earth, in contrast to the idealized geometrical figure of a reference ellipsoid.

The surface of the geoid is higher than the reference ellipsoid wherever there is a positive gravity anomaly (mass excess) and lower than the reference ellipsoid wherever there is a negative gravity anomaly (mass deficit).

Description

Deviation of the Geoid from the idealized figure of the Earth

(difference between the EGM96 geoid and the WGS84 reference ellipsoid)

Red areas are above the idealized ellipsoid; blue areas are below.

-107.0 m 0 m +85.4 m

Figure: *Map of the undulation of the geoid, in metres (based on the EGM96 gravity model and the WGS84 reference ellipsoid).*

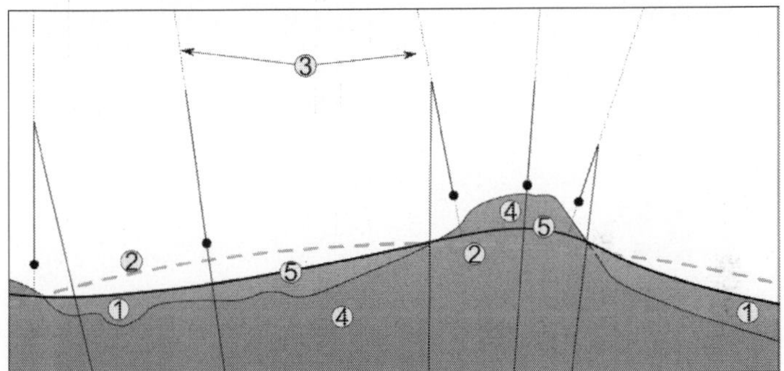

1. Ocean 2. Reference ellipsoid 3. Local plumb line 4. Continent 5. Geoid

The geoid surface is irregular, unlike the reference ellipsoid which is a mathematical idealized representation of the physical Earth, but considerably smoother than Earth's physical surface. Although the physical Earth has excursions of +8,000 m (Mount Everest) and –418 m (Dead Sea), the geoid's variation ranges from –106 to +85 m, less than 200 m total compared to a perfect mathematical ellipsoid.

If the ocean surface were isopycnic (of constant density) and undisturbed by tides, currents, or weather, it would closely approximate the geoid. The permanent deviation between the geoid and mean sea level is called ocean surface topography. If the continental land masses were criss-crossed by a series of tunnels or canals, the sea level in these canals would also very nearly coincide with the geoid. In reality the geoid does not have a physical meaning under the continents, but geodesists are able to derive the heights of continental points above this imaginary, yet physically defined, surface by a technique called spirit leveling.

Being an equipotential surface, the geoid is by definition a surface to which the force of gravity is everywhere perpendicular. This means that when travelling by ship, one does not notice the undulations of the geoid; the local vertical (plumb line) is always perpendicular to the geoid and the local horizon tangential to it. Likewise, spirit levels will always be parallel to the geoid.

Note that a GPS receiver on a ship may, during the course of a long voyage, indicate height variations, even though the ship will always be at sea level (tides not considered). This is because GPS satellites, orbiting about the centre of gravity of the Earth, can only measure heights relative to a geocentric reference ellipsoid. To obtain one's geoidal height, a raw GPS reading must be corrected. Conversely, height determined by spirit leveling from a tidal measurement station, as in traditional land surveying, will always be geoidal height. Modern GPS receivers have a grid implemented inside where they obtain the geoid (e.g. EGM-96) height over the World Geodetic System (WGS) ellipsoid from the current position. Then they are able to correct the height above WGS ellipsoid to the height above WGS84 geoid. In that case when the height is not zero on a ship it is due to various other factors such as ocean tides, atmospheric pressure (meteorological effects) and local sea surface topography.

Simplified Example

The gravitational field of the earth is neither perfect nor uniform. A flattened ellipsoid is typically used as the idealized earth, but even if the earth were perfectly spherical, the strength of gravity would not be the same everywhere, because density (and therefore mass) varies throughout the planet. This is due to magma distributions, mountain ranges, deep sea trenches, and so on.

If that perfect sphere were then covered in water, the water would not be the same height everywhere. Instead, the water level would be higher or lower depending on the particular strength of gravity in that location.

Spherical Harmonics Representation

Spherical harmonics are often used to approximate the shape of the geoid. The current best such set of spherical harmonic coefficients is EGM96 (Earth Gravity Model 1996), determined in an international collaborative project led by NIMA. The mathematical description of the non-rotating part of the potential function in this model is

$$V = \frac{GM}{r}\left(1 + \sum_{n=2}^{n_{\max}}\left(\frac{a}{r}\right)^n \sum_{m=0}^{n} \bar{P}_{nm}(\sin\phi)\left[\bar{C}_{nm}\cos m\lambda + \bar{S}_{nm}\sin m\lambda\right]\right),$$

where ϕ and λ are *geocentric* (spherical) latitude and longitude respectively, $\bar{P}_{nm}$ are the fully normalized associated Legendre polynomials of degree n and order m, and $\bar{C}_{nm}$ and $\bar{S}_{nm}$ are the numerical coefficients of the model based on measured data. Note that the above equation describes the Earth's gravitational potential V, not the geoid itself, at location $\phi, \lambda, r,$ the co-ordinate r being the *geocentric radius*, i.e., distance from the Earth's centre.

The geoid is a particular equipotential surface, and is somewhat involved to compute. The gradient of this potential also provides a model of the gravitational acceleration. EGM96 contains a full set of coefficients to degree and order 360 (i.e. $n_{\max} = 360$), describing details in the global geoid as small as 55 km (or 110 km, depending on your definition of resolution). The number of coefficients $\bar{C}_{nm}$, and $\bar{S}_{nm}$, can be determined by first observing in the equation for V that for a specific value of n there are two coefficients for every value of m except for m = 0. There is only one coefficient when m=0 since $\sin(0\lambda) = 0$. There are thus (2n+1) coefficients for every value of n. Using these facts and the formula, $\sum_{I=1}^{L} I = L(L+1)/2$, it follows that the total number of coefficients is given by

$\sum_{n=2}^{n_{\max}}(2n+1) = n_{\max}(n_{\max}+1) + n_{\max} - 3 = 130317$ using the EGM96 value of.

For many applications the complete series is unnecessarily complex and is truncated after a few (perhaps several dozen) terms.

New even higher resolution models are currently under development. For example, many of the authors of EGM96 are working on an updated model that should incorporate much of the new satellite gravity data, and should support up to degree and order 2160 (1/6 of a degree, requiring over 4 million coefficients). NGA has announced the availability of EGM2008, complete to spherical harmonic degree and order 2159, and contains additional coefficients extending to degree

2190 and order 2159. Software and data is on the Earth Gravitational Model 2008 (EGM2008) - WGS 84 Version page.

Precise Geoid

The 1990s saw important discoveries in the theory of geoid computation. The Precise Geoid Solution by Vaníθek and co-workers improved on the Stokesian approach to geoid computation. Their solution enables millimetre-to-centimetre accuracy in geoid computation, an order-of-magnitude improvement from previous classical solutions.

Variations in the height of the geoidal surface are related to density anomalous distributions within the Earth. Geoid measures help thus to understand the internal structure of the planet. Synthetic calculations show that the geoidal signature of a thickened crust (for example, in orogenic belts produced by continental collision) is positive, opposite to what should be expected if the thickening affects the entire lithosphere.

Time-Variability

Recent satellite missions, such as GOCE and GRACE, have enabled the study of time-variable geoid signals. The first products based on GOCE satellite data became available online in June 2010, through the European Space Agency (ESA)'s Earth observation user services tools. ESA launched the satellite in March 2009 on a mission to map Earth's gravity with unprecedented accuracy and spatial resolution. On 31 March 2011, the new geoid model was unveiled at the Fourth International GOCE User Workshop hosted at the Technische Universität München in Munich, Germany. Studies using the time-variable geoid computed from GRACE data have provided information on global hydrologic cycles, mass balances of ice sheets, and postglacial rebound. From postglacial rebound measurements, time-variable GRACE data can be used to deduce the viscosity of Earth's mantle.

Chapter 4

Magnetism

Magnetism is a class of physical phenomena that are mediated by magnetic fields. Electric currents and the fundamental magnetic moments of elementary particles give rise to a magnetic field, which acts on other currents and magnetic moments. All materials are influenced to some extent by a magnetic field. The most familiar effect is on permanent magnets, which have persistent magnetic moments caused by ferromagnetism. Most materials do not have permanent moments. Some are attracted to a magnetic field (paramagnetism); others are repulsed by a magnetic field (diamagnetism); others have a much more complex relationship with an applied magnetic field (spin glass behaviour and antiferromagnetism). Substances that are negligibly affected by magnetic fields are known as *non-magnetic* substances. They include copper, aluminium, gases, and plastic. Pure oxygen exhibits magnetic properties when cooled to a liquid state.

The magnetic state (or phase) of a material depends on temperature (and other variables such as pressure and the applied magnetic field) so that a material may exhibit more than one form of magnetism depending on its temperature, etc.

Diamagnetism

Diamagnetism appears in all materials, and is the tendency of a material to oppose an applied magnetic field, and therefore, to be repelled by a magnetic field. However, in a material with paramagnetic properties (that is, with a tendency to enhance an external magnetic field), the paramagnetic behaviour dominates. Thus, despite its universal occurrence, diamagnetic behaviour is observed only in a purely diamagnetic material. In a diamagnetic material, there are no unpaired electrons, so the intrinsic electron magnetic moments cannot produce any bulk effect. In these cases, the magnetization arises from the electrons' orbital motions, which can be understood classically as follows:

When a material is put in a magnetic field, the electrons circling the nucleus will experience, in addition to their Coulomb attraction to the nucleus, a Lorentz force from the magnetic field. Depending on which direction the electron is orbiting, this force may increase the centripetal force on the electrons, pulling them in towards the nucleus, or it may decrease the force, pulling them away from the nucleus. This effect systematically increases the orbital magnetic moments that were aligned opposite the field, and decreases the ones aligned parallel to the field (in accordance with Lenz's law). This results in a small bulk magnetic moment, with an opposite direction to the applied field.

Note that this description is meant only as an heuristic; a proper understanding requires a quantum-mechanical description.

Note that all materials undergo this orbital response. However, in paramagnetic and ferromagnetic substances, the diamagnetic effect is overwhelmed by the much stronger effects caused by the unpaired electrons.

Paramagnetism

In a paramagnetic material there are *unpaired electrons*, i.e. atomic or molecular orbitals with exactly one electron in them. While paired electrons are required by the Pauli exclusion principle to have their intrinsic ('spin') magnetic moments pointing in opposite directions, causing their magnetic fields to cancel out, an unpaired electron is free to align its magnetic moment in any direction. When an external magnetic field is applied, these magnetic moments will tend to align themselves in the same direction as the applied field, thus reinforcing it.

Ferromagnetism

A ferromagnet, like a paramagnetic substance, has unpaired electrons. However, in *addition* to the electrons' intrinsic magnetic moment's tendency to be parallel to an *applied field*, there is also in these materials a tendency for these magnetic moments to orient parallel to *each other* to maintain a lowered-energy state. Thus, even in the absence of an applied field, the magnetic moments of the electrons in the material spontaneously line up parallel to one another.

Every ferromagnetic substance has its own individual temperature, called the Curie temperature, or Curie point, above which it loses its ferromagnetic properties. This is because the thermal tendency to disorder overwhelms the energy-lowering due to ferromagnetic order.

Ferromagnetism only occurs in a few substances; the common ones are iron, nickel, cobalt, their alloys, and some alloys of rare earth metals.

Magnetic Dipoles

A very common source of magnetic field shown in nature is a dipole, with a "South pole" and a "North pole", terms dating back to the use of magnets as compasses, interacting with the Earth's magnetic field to indicate North and South on the globe. Since opposite ends of magnets are attracted, the north pole of a magnet is attracted to the south pole of another magnet. The Earth's North Magnetic Pole (currently in the Arctic Ocean, north of Canada) is physically a south pole, as it attracts the north pole of a compass. A magnetic field contains energy, and physical systems move toward configurations with lower energy. When diamagnetic material is placed in a magnetic field, a *magnetic dipole* tends to align itself in opposed polarity to that field, thereby lowering the net field strength. When ferromagnetic material is placed within a magnetic field, the magnetic dipoles align to the applied field, thus expanding the domain walls of the magnetic domains.

Magnetic Monopoles

Since a bar magnet gets its ferromagnetism from electrons distributed evenly throughout the bar, when a bar magnet is cut in half, each of the resulting pieces is a smaller bar magnet. Even though a magnet is said to have a north pole and a south pole, these two poles cannot be separated from each other. A monopole—if such a thing exists—would be a new and fundamentally different kind of magnetic object. It would act as an isolated north pole, not attached to a south pole, or vice versa. Monopoles would carry "magnetic charge" analogous to electric charge. Despite systematic searches since 1931, as of 2010, they have never been observed, and could very well not exist.

Nevertheless, some theoretical physics models predict the existence of these magnetic monopoles. Paul Dirac observed in 1931 that, because electricity and magnetism show a certain symmetry, just as quantum theory predicts that individual positive or negative electric charges can be observed without the opposing charge, isolated South or North magnetic poles should be observable. Using quantum theory Dirac showed that if magnetic monopoles exist, then one could explain the quantization of electric charge—that is, why the observed elementary particles carry charges that are multiples of the charge of the electron.

Certain grand unified theories predict the existence of monopoles which, unlike elementary particles, are solitons (localized energy packets). The initial results of using these models to estimate the number of monopoles created in the big bang contradicted cosmological observations—the monopoles would have been so plentiful and massive that they would have long since halted the expansion of the universe. However, the idea of inflation (for which this problem served as a partial motivation) was successful in solving this problem, creating models in which monopoles existed but were rare enough to be consistent with current observations.

Quantum-Mechanical Origin of Magnetism

In principle all kinds of magnetism originate (similar to Superconductivity) from specific quantum-mechanical phenomena (e.g. Mathematical formulation of quantum mechanics, in particular the chapters on spin and on the Pauli principle).

A successful model was developed already in 1927, by Walter Heitler and Fritz London, who derived quantum-mechanically, how hydrogen molecules are formed from hydrogen atoms, i.e. from the atomic hydrogen orbitals u_A and u_B centred at the nuclei A and B. That this leads to magnetism is not at all obvious, but will be explained in the following.

According to the Heitler-London theory, so-called two-body molecular σ-orbitals are formed, namely the resulting orbital is:

$$\psi(\mathbf{r}_1,\mathbf{r}_2)=\frac{1}{\sqrt{2}}\left(u_A(\mathbf{r}_1)u_B(\mathbf{r}_2)+u_B(\mathbf{r}_1)u_A(\mathbf{r}_2)\right)$$

Here the last product means that a first electron, r_1, is in an atomic hydrogen-orbital centred at the second nucleus, whereas the second electron runs around the first nucleus. This “exchange” phenomenon is an expression for the quantum-mechanical property that particles with identical properties cannot be distinguished. It is specific not only for the formation of chemical bonds, but as we will see, also for magnetism, i.e. in this connection the term exchange interaction arises, a term which is essential for the origin of magnetism, and which is stronger, roughly by factors 100 and even by 1000, than the energies arising from the electrodynamic dipole-dipole interaction.

As for the *spin function* $\chi(s_1,s_2)$, which is responsible for the magnetism, we have the already mentioned Pauli's principle, namely that a symmetric orbital (i.e. with the + sign as above) must be multiplied with an antisymmetric spin function (i.e. with a “ sign), and *vice versa*. Thus:

$$\chi(s_1,s_2)=\frac{1}{\sqrt{2}}\left(\alpha(s_1)\beta(s_2)-\beta(s_1)\alpha(s_2)\right)_{,,}$$

I.e., not only u_A and u_B must be substituted by α and β, respectively (the first entity means “spin up”, the second one “spin down”), but also the sign + by the “ sign, and finally r_i by the discrete values s_i (= ±½); thereby we have $\alpha(+1/2)=\beta(-1/2)=1$ and $\alpha(-1/2)=\beta(+1/2)=0$. The “singlet state”, i.e. the “ sign, means: the spins are *antiparallel*, i.e. for the solid we have antiferromagnetism, and for two-atomic molecules one has diamagnetism.

The tendency to form a (homoeopolar) chemical bond (this means: the formation of a *symmetric* molecular orbital, i.e. with the + sign) results through the Pauli principle automatically in an *antisymmetric* spin state (i.e. with the “ sign). In contrast, the Coulomb repulsion of the electrons, i.e. the tendency that they try to avoid each other by this repulsion, would lead to an *antisymmetric* orbital function (i.e. with the “ sign) of these two particles, and complementary to a *symmetric* spin function (i.e. with the + sign, one of the so-called “triplet functions”). Thus, now the spins would be *parallel* (ferromagnetism in a solid, paramagnetism in two-atomic gases).

The last-mentioned tendency dominates in the metals iron, cobalt and nickel, and in some rare earths, which are *ferromagnetic*. Most of the other metals, where the first-mentioned tendency dominates, are *nonmagnetic* (e.g. sodium, aluminium, and magnesium) or *antiferromagnetic* (e.g. manganese). Diatomic gases are also almost exclusively diamagnetic, and not paramagnetic. However, the oxygen molecule, because of the involvement of π-orbitals, is an exception important for the life-sciences.

The Heitler-London considerations can be generalized to the Heisenberg model of magnetism.

The explanation of the phenomena is thus essentially based on all subtleties of quantum mechanics, whereas the electrodynamics covers mainly the phenomenology.

Geomagnetism

Since the ancient times, people have known that certain materials would always point to a fixed direction. Ancient Chinese, based on this principle, invented compass (known as

Si-nan at the time). These materials behave in this way because they are magnetized and the Earth is surrounded by a magnetic field. Whichever direction such objects are turned to, they would always revert to the direction of the magnetic field, which is known as the Earth's geomagnetic field.

Cause the Earth's Geomagnetic Field

The Earth's geomagnetic field is a combination of several magnetic fields generated by various sources superimposing on each other. More than 90% of the field is generated by the movement of conducting material inside the Earth's core, which is often referred to as the Main Field. Other important sources of the geomagnetic field include electric current flowing in the ionized upper atmosphere and currents flowing within the earth's crust. There are also local anomalies produced by mountain ranges, ore deposits, geological faults, and artificial products such as trains, aircraft, power lines etc.

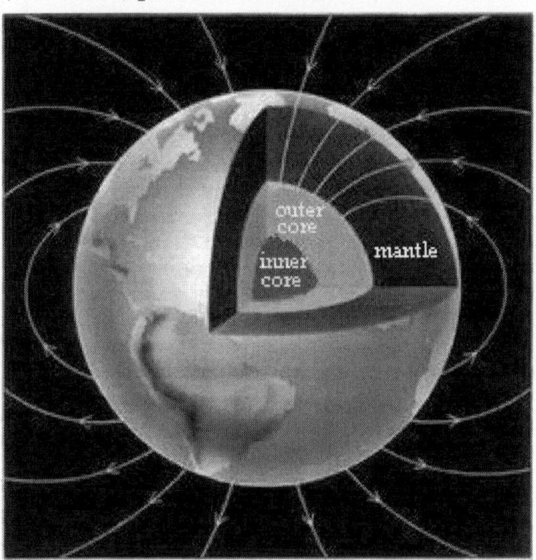

Figure: *The Earth's magnetic field is caused primarily by conducting material inside the Earth's core*

Effect on Human Beings and Other Life Forms

The Sun is continuously ejecting high energy charged particles (known as solar wind) which, if ever able to reach the Earth's surface directly, could harm human beings and other life forms on Earth. Fortunately, the Earth's geomagnetic field acts as a shield and protects life on Earth from solar winds. Besides, some animals, such as sea turtles and migratory birds, rely on the Earth's geomagnetic field for navigation.

Figure: *The Earth's geomagnetic field shields the Earth from direct impact of the solar wind.*

Application of Geomagnetic Knowledge

The discovery of geomagnetism and the invention of compass enabled seafaring and paved the way for the Age of Exploration, causing tremendous impacts on trades, wars and cultural exchanges.

To date, compass is still a necessary tool for activities such as hiking, stargazing as well as marine and air navigation.

Geomagnetic Directions

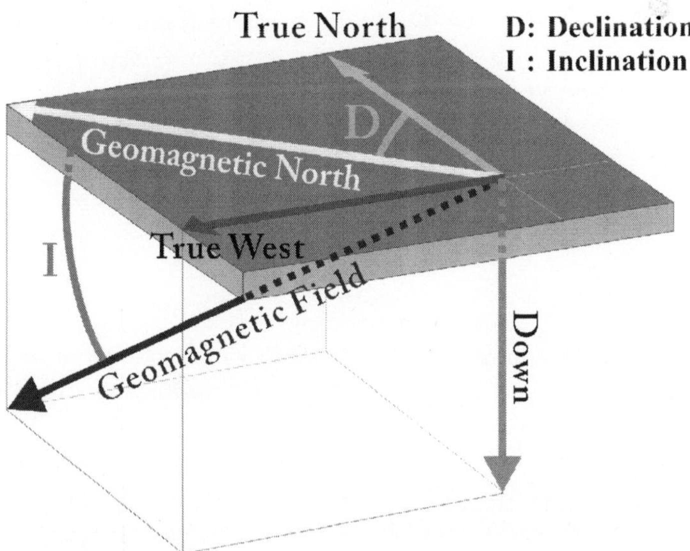

Figure: *Geomagnetic declination and inclination*

Although a compass points generally to the north and south, careful examination reveals that the north indicated on a compass (magnetic north) differs from the "north" in the geographical sense (true north, as determined by observing the positions of the sun or other stars). The difference is known as geomagnetic declination. Besides, the angle between the geomagnetic field and the horizontal plane is known as geomagnetic inclination.

Rock Magnetism

Rock magnetism is the study of the magnetic properties of rocks, sediments and soils. The field arose out of the need in paleomagnetism to understand how rocks record the Earth's magnetic field. This remanence is carried by minerals, particularly certain strongly magnetic minerals like magnetite (the main source of magnetism in lodestone). An understanding of remanence helps paleomagnetists to develop methods for measuring the ancient magnetic field and correct for effects like sediment compaction and metamorphism. Rock magnetic methods are used to get a more detailed picture of the source of distinctive striped pattern in marine magnetic anomalies that provides important information on plate tectonics. They are also used to interpret terrestrial magnetic anomalies in magnetic surveys as well as the strong crustal magnetism on Mars.

Strongly magnetic minerals have properties that depend on the size, shape, defect structure and concentration of the minerals in a rock. Rock magnetism provides non-destructive methods for analyzing these minerals such as magnetic hysteresis measurements, temperature-dependent remanence measurements, Mössbauer spectroscopy, ferromagnetic resonance and so on. With such methods, rock magnetists can measure the effects of past climate change and human impacts on the mineralogy. In sediments, a lot of the magnetic remanence is carried by minerals that were created by magnetotactic bacteria, so rock magnetists have made significant contributions to biomagnetism.

History

Until the 20th century, the study of the Earth's field (geomagnetism and paleomagnetism) and of magnetic materials (especially ferromagnetism) developed separately.

Rock magnetism had its start when scientists brought these two fields together in the laboratory. Koenigsberger (1938), Thellier (1938) and Nagata (1943) investigated the origin of remanence in igneous rocks. By heating rocks and archeological materials to high temperatures in a magnetic field, they gave the materials a thermoremanent magnetization (TRM), and they investigated the properties of this magnetization. Thellier developed a series of conditions (the Thellier laws) that, if fulfilled, would allow the determination of the intensity of the ancient magnetic field to be determined using the Thellier-Thellier method. In 1949, Louis Néel developed a theory that explained these observations, showed that the Thellier laws were satisfied by certain kinds of single-domain magnets, and introduced the concept of blocking of TRM.

When paleomagnetic work in the 1950s lent support to the theory of continental drift, skeptics were quick to question whether rocks could carry a stable remanence for geological ages. Rock magnetists were able to show that rocks could have more than one component of remanence, some soft (easily removed) and some very stable. To get at the stable part, they took to "cleaning" samples by heating them or exposing them to an alternating field.

However, later events, particularly the recognition that many North American rocks had been pervasively remagnetized in the Paleozoic, showed that a single cleaning step was inadequate, and paleomagnetists began to routinely use stepwise demagnetization to strip away the remanence in small bits.

Fundamentals

Types of Magnetic Order

The contribution of a mineral to the total magnetism of a rock depends strongly on the type of magnetic order or disorder. Magnetically disordered minerals (diamagnets and paramagnets) contribute a weak magnetism and have no remanence. The more important minerals for rock magnetism are the minerals that can be magnetically ordered, at least at some temperatures. These are the ferromagnets, ferrimagnets and certain kinds of antiferromagnets. These minerals have a much stronger response to the field and can have a remanence.

Diamagnetism

Diamagnetism is a magnetic response shared by all substances. In response to an applied magnetic field, electrons precess, and by Lenz's law they act to shield the interior of a body from the magnetic field. Thus, the moment produced is in the opposite direction to the field and the susceptibility is negative. This effect is weak but independent of temperature.

A substance whose only magnetic response is diamagnetism is called a diamagnet.

Paramagnetism

Paramagnetism is a weak positive response to a magnetic field due to rotation of electron spins. Paramagnetism occurs in certain kinds of iron-bearing minerals because the iron contains an unpaired electron in one of their shells. Some are paramagnetic down to absolute zero and their susceptibility is inversely proportional to the temperature; others are magnetically ordered below a critical temperature and the susceptibility increases as it approaches that temperature.

Ferromagnetism

Collectively, strongly magnetic materials are often referred to as ferromagnets. However, this magnetism can arise as the result of more than one kind of magnetic order. In the strict sense, ferromagnetism refers to magnetic ordering where neighbouring electron spins are aligned by the exchange interaction. The classic ferromagnet is iron. Below a critical temperature called the Curie temperature, ferromagnets have a spontaneous magnetization and there is hysteresis in their response to a changing magnetic field. Most importantly for rock magnetism, they have remanence, so they can record the Earth's field. Iron does not occur widely in its pure form. It is usually incorporated into iron oxides, oxyhydroxides and sulfides. In these compounds, the iron atoms are not close enough for direct exchange, so they are coupled by indirect exchange or superexchange. The result is that the crystal lattice is divided into two or more sublattices with different moments.

Ferrimagnetism

Ferrimagnets have two sublattices with opposing moments. One sublattice has a larger moment, so there is a net unbalance. Magnetite, the most important of the magnetic minerals, is a ferrimagnet. Ferrimagnets often behave like ferromagnets, but the temperature dependence of their spontaneous magnetization can be quite different. Louis Néel identified four types of temperature dependence, one of which involves a reversal of the magnetization. This phenomenon played a role in controversies over marine magnetic anomalies.

Antiferromagnetism

Antiferromagnets, like ferrimagnets, have two sublattices with opposing moments, but now the moments are equal in magnitude. If the moments are exactly opposed, the magnet has no remanence. However, the moments can be tilted (spin canting), resulting in a moment nearly at right angles to the moments of the sublattices. Hematite has this kind of magnetism.

Magnetic Mineralogy

Magnetic mineralogy is the study of the magnetic properties of minerals. The contribution of a mineral to the total magnetism of a rock depends strongly on the type of magnetic order or disorder. Magnetically disordered minerals (diamagnets and paramagnets) contribute a weak magnetism and have no remanence. The more important minerals for rock magnetism are the minerals that can be magnetically ordered, at least at some temperatures. These are the ferromagnets, ferrimagnets and certain kinds of antiferromagnets. These minerals have a much stronger response to the field and can have a remanence.

Types of Remanence

Magnetic remanence is often identified with a particular kind of remanence that is obtained after exposing a magnet to a field at room temperature. However, the Earth's field is not large, and this kind of remanence would be weak and easily overwritten by later fields. A central part of rock magnetism is the study of magnetic remanence, both as natural remanent magnetization (NRM) in rocks obtained from the field and remanence induced in the laboratory. Below are listed the important natural remanences and some artificially induced kinds.

Thermoremanent Magnetization (TRM)

When an igneous rock cools, it acquires a *thermoremanent magnetization (TRM)* from the Earth's field. TRM can be much larger than it would be if exposed to the same field at room temperature. This remanence can also be very stable, lasting without significant change for millions of years. TRM is the main reason that paleomagnetists are able to deduce the direction and magnitude of the ancient Earth's field.

If a rock is later re-heated (as a result of burial, for example), part or all of the TRM can be replaced by a new remanence. If it is only part of the remanence, it is known as *partial thermoremanent magnetization (pTRM)*. Because numerous experiments have been done modelling different ways of acquiring remanence, pTRM can have other meanings. For example, it can also be acquired in the laboratory by cooling in zero field to a temperature T_1 (below the Curie temperature), applying a magnetic field and cooling to a temperature T_2, then cooling the rest of the way to room temperature in zero field.

The standard model for TRM is as follows. When a mineral such as magnetite cools below the Curie temperature, it becomes ferromagnetic but is not immediately capable of carrying a remanence. Instead, it is superparamagnetic, responding reversibly to changes in the magnetic field. For remanence to be possible there must be a strong enough magnetic anisotropy to keep the magnetization near a stable state; otherwise, thermal fluctuations make the magnetic moment wander randomly. As the rock continues to cool, there is a critical temperature at which the magnetic anisotropy becomes large enough to keep the moment from wandering: this temperature is called the *blocking temperature* and referred to by the symbol T_B. The magnetization remains in the same state as the rock is cooled to room temperature and becomes a thermoremanent magnetization.

Chemical (or Crystallization) Remanent Magnetization (CRM)

Magnetic grains may precipitate from a circulating solution, or be formed during chemical reactions, and may record the direction of the magnetic field at the time of mineral formation. The field is said to be recorded by *chemical remanent magnetization (CRM).* The mineral recording the field commonly is hematite, another iron oxide. Redbeds, clastic sedimentary rocks (such as sandstones) that are red primarily because of hematite formation during or after sedimentary diagenesis, may have useful CRM signatures, and magnetostratigraphy can be based on such signatures.

Depositional Remanent Magnetization (DRM)

Magnetic grains in sediments may align with the magnetic field during or soon after deposition; this is known as detrital remnant magnetization (DRM). If the magnetization is acquired as the grains are deposited, the result is a depositional detrital remanent magnetization (dDRM); if it is acquired soon after deposition, it is a *post-depositional detrital remanent magnetization (pDRM).*

Viscous Remanent Magnetization

Viscous remanent magnetization (VRM), also known as viscous magnetization, is remanence that is acquired by ferromagnetic minerals by sitting in a magnetic field for some time. The natural remanent magnetization of an igneous rock can be altered by this process. To remove this component, some form of stepwise demagnetization must be used.

Environmental Magnetism

Environmental magnetism is the study of magnetism as it relates to the effects of climate, sediment transport, pollution and other environmental influences on magnetic minerals. It makes use of techniques from rock magnetism and magnetic mineralogy. The magnetic properties of minerals are used as proxies for environmental change in applications such as paleoclimate, paleoceanography, studies of the provenance of sediments, pollution and archeology. The main advantages of using magnetic measurements are that magnetic minerals are almost ubiquitous and magnetic measurements are quick and non-invasive.

Environmental magnetism was first identified as a distinct field in 1978 and was introduced to a wider audience by the book *Environmental Magnetism* in 1986. Since then it has grown rapidly, finding application in and making major contributions to a range of diverse fields, especially paleoclimate, sedimentology, paleoceanography, and studies of particulate pollution.

Fundamentals

Environmental magnetism is built on two parts of rock magnetism: magnetic mineralogy, which looks at how basic magnetic properties depend on composition; and magnetic hysteresis, which can provide details on particle size and other physical properties that also affect the hysteresis. Several parameters such as magnetic susceptibility and various kinds of remanence have been developed to represent certain features of the hysteresis. These parameters are then used to estimate mineral size and composition. The main contributors to the magnetic properties of rocks are the iron oxides, including magnetite, maghemite, hematite; and iron sulfides (particularly greigite and pyrrhotite). These minerals are strongly magnetic because, at room temperature, they are magnetically ordered (magnetite, maghemite and greigite are ferrimagnets while hematite is a canted antiferromagnet).

To relate magnetic measurements to the environment, environmental magnetists have identified a variety of processes that give rise to each magnetic mineral. These include erosion, transport, fossil fuel combustion, and bacterial formation. The latter includes extracellular precipitation and formation of magnetosomes by magnetotactic bacteria.

Applications

Paleoclimate: Magnetic measurements have been used to investigate past climate. A classic example is the study of loess, which is windblown dust from the edges of glaciers and semiarid desert margins. In north-central China, blankets of loess that were deposited during glacial periods alternate with paleosols (fossil soils) that formed during warmer and wetter interglacials. The magnetic susceptibility profiles of these sediments have been dated using magnetostratigraphy, which identifies geomagnetic reversals, and correlated with climate indicators such as oxygen isotope stages. Ultimately, this work allowed environmental magnetists to map out the variations in the monsoon cycle during the Quaternary.

Biomagnetism

Biomagnetism is the phenomenon of magnetic fields *produced* by living organisms; it is a subset of bioelectromagnetism. In contrast, organisms' use of magnetism in navigation is magnetoception and the study of the magnetic fields' *effects* on organisms is *magnetobiology*. (The word biomagnetism has also been used loosely to include magnetobiology, further encompassing almost any combination of the words magnetism, cosmology, and biology, such as "magnetoastrobiology".)

The origin of the word biomagnetism is unclear, but seems to have appeared several hundred years ago, linked to the expression "animal magnetism." The present scientific definition took form in the 1970s, when an increasing number of researchers began to measure the magnetic fields produced by the human body. The first valid measurement was actually made in 1963, but the field began to expand only after a low-noise technique was developed in 1970. Today the community of biomagnetic researchers does not have a formal organisation, but international conferences are held every two years, with about 600 attendees. Most conference activity centres on the MEG (magnetoencephalogram), the measurement of the magnetic field of the brain.

In 2013 a study published in the journal Frontiers in Zoology found that "dogs preferred to excrete with the body being aligned along the North–south axis under calm MF (magnetic field) conditions", and that these findings "open new horizons for biomagnetic research."

Magnetoception

Magnetoception (or magnetoreception as it was first referred to in 1972) is a sense which allows an organism to detect a magnetic field to perceive direction, altitude or location. This sense has been proposed to explain animal navigation in vertebrates and insects, and as a method for animals to develop regional maps. For the purpose of navigation, magnetoception deals with the detection of the Earth's magnetic field.

Magnetoception has been observed in bacteria, in invertebrates such as fruit flies, lobsters and honeybees. It has also been demonstrated in vertebrates including birds, turtles, sharks and stingrays. Magnetoception in humans is controversial.

Proposed Mechanisms

An unequivocal demonstration of the use of magnetic fields for orientation within an organism has been in a class of bacteria known as magnetotactic bacteria. These bacteria demonstrate a behavioural phenomenon known as magnetotaxis, in which the bacterium orients itself and migrates in the direction along the Earth's magnetic field lines. The bacteria contain magnetosomes, which are particles of magnetite or iron sulfide enclosed within the bacterial cells. Each bacterium cell essentially acts as a magnetic dipole. They form in chains where the moments of each magnetosome align in parallel, giving the bacteria its permanent-magnet characteristics. These chains are formed symmetrically to preserve the crystalline structure of the cells. These bacteria are said to have permanent magnetic sensitivity.

For animals the mechanism for magnetoception is unknown, but there exist two main hypotheses to explain the phenomenon. According to one model, cryptochrome, when exposed to blue light, becomes activated to form a pair of two radicals (molecules with a single unpaired electron) where the spins of the two unpaired electrons are correlated. The surrounding magnetic field affects the kind of this correlation (parallel or anti-parallel), and this in turn affects the length of time cryptochrome stays in its activated state. Activation of cryptochrome may affect the light-sensitivity of retinal neurons, with the overall result that the bird can "see" the magnetic field. The Earth's magnetic field is only 0.5 Gauss and so it is difficult to conceive of a mechanism by which such a field could lead to any chemical changes other than those affecting the weak magnetic fields between radical pairs. Cryptochromes are therefore thought to be essential for the light-dependent ability of the fruit fly *Drosophila melanogaster* to sense magnetic fields.

The second proposed model for magnetoreception relies on Fe_3O_4, also referred to as iron (II, III) oxide or magnetite, a natural oxide with strong magnetism. Iron (II, III) oxide remains permanently magnetized when its length is larger than 50 nm and becomes magnetized when exposed to a magnetic field if its length is less than 50 nm. In both of these situations the Earth's magnetic field leads to a transducible signal via a physical effect on this magnetically sensitive oxide.

Another less general type of magnetic sensing mechanism in animals that has been thoroughly described is the inductive sensing methods used by sharks, stingrays and chimaeras (cartilaginous fish). These species possess a unique electroreceptive organ known as *ampullae of Lorenzini* which can detect a slight variation in electric potential. These organs are made up of mucus-filled canals that connect from the skin's pores to small sacs within the animal's flesh that are also filled with mucus. The ampullae of Lorenzini are capable of detecting DC currents and have been proposed to be used in the sensing of the weak electric fields of prey and predators. These organs could also possibly sense magnetic fields, by means of Faraday's law: as a conductor moves through a magnetic field an electric potential is generated. In this case the conductor is the animal moving through a magnetic field, and the potential induced depends on the time varying rate of flux through the conductor according to

$$V_{ind} = -\frac{d\phi}{dt}.$$

These organs detect very small fluctuations in the potential difference between the pore and the base of the electroreceptor sack. An increase in potential results in a decrease in the rate of nerve activity, and a decrease in potential results in an increase in the rate of nerve activity. This is analogous to the behaviour of a current carrying conductor; with a fixed channel resistance, an increase in potential would decrease the amount of current detected, and vice versa. These receptors are located along the mouth and nose of sharks and stingrays.

In Invertebrates

The mollusc *Tochuina tetraquetra* (formerly *Tritonia diomedea* or *Tritonia gigantea*) has been studied for clues as to the neural mechanism behind magnetoreception in a species. Some of the earliest work with *Tochuina* showed that prior to a full moon *Tochuina* would orient their bodies between magnetic north and east. A Y-maze was established with a right turn equal to geomagnetic south and a left turn equal to geomagnetic east. Within this geomagnetic field 80% of *Tochuina* made a turn to the left or magnetic east.

However, when a reversed magnetic field was applied that rotated magnetic north 180° there was no significant preference for either turn, which now corresponded with magnetic north and magnetic west. These results, though interesting, do not conclusively establish that *Tochuina* uses magnetic fields in magnetoreception. These experiments do not include a control for the activation of the Rubens' coil in the reversed magnetic field experiments. Therefore, it is possible that heat or noise generated by the coil was responsible for the loss of choice preference.

Further work with *Tochuina* was unable to identify any neurons that showed rapid changes in firing as a result of magnetic fields. However, pedal 5 neurons, two bisymmetric neurons located within the *Tochuina* pedal ganglion, exhibited gradual changes in firing over time following 30 minutes of magnetic stimulation provided by a Rubens' coil. Further studies showed that pedal 7 neurons in the pedal ganglion were inhibited when exposed to magnetic fields over the course of 30 minutes. The function of both pedal 5 neurons and pedal 7 neurons is currently unknown.

Drosophila melanogaster is another invertebrate which may be able to orient to magnetic fields. Experimental techniques such as gene knockouts have allowed a closer examination of possible magnetoreception in these fruit flies. Various *Drosophila* strains have been trained to respond to magnetic fields. In a choice test flies were loaded into an apparatus with two arms that were surrounded by electric coils. Current was run through each of the coils, but only one was configured to produce a 5-Gauss magnetic field at a time. The flies in this T-maze were tested on their native ability to recognise the presence of the magnetic field in an arm and on their response following training where the magnetic field was paired with a sucrose reward. Many of the strains of flies showed a learned preference for the magnetic field following training.

However, when the only cryptochrome found in *Drosophila*, type 1 Cry, is altered, either through a missense mutation or replacement of the Cry gene, the flies exhibit a loss of magnetosensitivity. Furthermore, when light is filtered to only allow wavelengths greater than 420 nm through, *Drosophila* loses its trained response to magnetic fields. This response to filtered light is likely linked to the action spectrum of fly-cryptochrome which has a range from 350 nm – 400 nm and plateaus from 430-450 nm. Although researchers had believed that a tryptophan triad in cryptochrome was responsible for the free radicals on which magnetic fields could act, recent work with *Drosophila* has shown that tryptophan might not be behind cryptochrome dependent magnetoreception.

Alteration of the tryptophan protein does not result in the loss of magnetosensitivity of a fly expressing either type 1 Cry or the cryptochrome found in vertebrates, type 2 Cry. Therefore it remains unclear exactly how cryptochrome mediates magnetoreception. These experiments used a 5 gauss magnetic field, 10 times the strength of the Earth's magnetic field). *Drosophila* has not been shown to be able to respond to the Earth's weaker magnetic field.

In Homing Pigeons

Homing pigeons have been known to use magnetic fields as part of their complex navigation system. Work by William Keeton showed that homing pigeons that were time shifted were unable to orient themselves correctly on a clear sunny day. This was considered a result of the fact that homing pigeons who used the sun for navigation would have to compensate for its movement throughout the day and a time shifted pigeon would be incapable of doing such compensation properly. However, if time shifted pigeons were released on overcast day they navigated correctly. This led to the hypothesis that under particular conditions homing pigeons rely on magnetic fields to orient themselves. Further experiments with magnets attached to the backs of homing pigeons demonstrated that disruption of the bird's ability to sense the Earth's magnetic field leads to a loss of proper orientation behaviour under overcast conditions.

There have been two mechanisms implicated in homing pigeon magnetoreception : the visually mediated free-radical pair mechanism and a magnetite based directional compass or inclination compass. More recent behavioural tests have shown that pigeons are able to detect magnetic anomalies of 186 microtesla (1.86 Gauss).

In a choice test birds were trained to jump onto a platform on one end of a tunnel if there was no magnetic field present and to jump onto a platform on the other end of the tunnel if a magnetic field was present. In this test, birds were rewarded with a food prize and punished with a time penalty. Homing pigeons were able to make the correct choice 55%-65% of the time which is higher than what would be expected if the pigeons were simply guessing. The ability of pigeons to detect a magnetic field is impaired by application of ligocaine, an anesthetic, to the olfactory mucosa. Furthermore, sectioning the trigeminal nerve leads to an inability to detect a magnetic field, while sectioning of the olfactory nerve has no effect on the magnetic sense of homing pigeons. These results suggest that magnetite located in the beak of pigeons may be responsible for magnetoreception via trigeminal mediation. However, it has not been shown that the magnetite located in the beak of pigeons is capable of responding to a magnetic field with the Earth's strength. Therefore the receptor responsible for magnetosensitivity in homing pigeons has not been cemented.

Aside from the sensory receptor for magnetic reception in homing pigeons there has been work on neural regions that are possibly involved in the processing of magnetic information within the brain. Areas of the brain that have shown increases in activity in response to magnetic fields with a strength of 50 or 150 microtesla are the posterior vestibular nuclei, dorsal thalamus, hippocampus, and visual hyperpallium.

As previously mentioned, pigeons provided some of the first evidence for the use of magnetoreception in navigation. As a result, they have been an organism of focus in magnetoreception studies. The precise mechanism used by pigeons has not been established and so it is unclear yet whether pigeons rely solely on a cryptochrome-mediated receptor or on beak-magnetite.

In Domestic Hens

Domestic hens have iron mineral deposits in the dendrites in the upper beak and are capable of magnetoreception. Because hens use directional information from the magnetic field of the earth to orient in relatively small areas, this raises the possibility that beak-trimming (removal of part of the beak to reduce injurious pecking frequently performed on egg-laying hens) impairs the ability of hens to orient in extensive systems, or to move in and out of buildings in free-range systems.

In Mammals

Work with mice, mole-rats and bats has shown that some mammals are capable of magnetoception. When woodmice are removed from their home area and deprived of visual and olfactory cues, they orient themselves towards their homes until an inverted magnetic field is applied to their cage. However, when the same mice are allowed access to visual cues, they are able to orient themselves towards home despite the presence of inverted magnetic fields. This indicates that when woodmice are displaced, they use magnetic fields to orient themselves if there are no other cues available. However, studies such as this have been criticized because of the difficulty of completely removing sensory cues and the fact that in some of these studies, the magnetic field is artificially changed before the test rather than during the test. Due to the timing of activating the magnetic fields, the results of these experiments do not conclusively show that woodmice respond to magnetic fields when deprived of other cues.

Research with the Zambian mole-rat, a subterranean mammal, has led to reports that they use magnetic fields as a polarity compass to aid in the orientation of their nests. In contrast to work with woodmice, Zambian mole-rats do not exhibit different orientation behaviour when a visual cue such as the sun is present, a result that has been suggested is due to their subterranean lifestyle. Further investigation of mole-rat magnetoreception lead to the finding that exposure to magnetic fields leads to an increase in neural activity within the superior colliculus as measured by immediate early gene expression. The activity level of neurons within two levels of the superior colliculus, the outer sublayer of the intermediate gray layer and the deep gray layer, were elevated in a non-specific manner when exposed to various magnetic fields. However, within the inner sublayer of the intermediate gray layer (InGi) there were two or three clusters of responsive cells. The more time the mole rats were exposed to a magnetic field the greater the immediate early gene expression within the InGi. However, if Zambian mole-rats were placed in a field with a shielded magnetic field only a few scattered cells were active. Therefore it has been proposed that in mammals, the superior colliculus is an important neural structure in the processing of magnetic information.

Bats may also use magnetic fields to orient themselves. While it is known that bats use echolocation to navigate over short distances, it is unclear how they navigate over longer distances. When *Eptesicus fuscus* are taken from their home roosts and exposed to magnetic fields 90 degrees clockwise or counterclockwise of magnetic north, they are disoriented and set off for their homes in the wrong direction. Therefore, it seems that *Eptesicus fuscus* is capable of magnetoreception. However, the exact use of magnetic fields by *Eptesicus fuscus* is unclear as the magnetic field could be being used either as a map, a compass, or a compass calibrator. Recent research with another bat species, *Myotis myotis*, supports the hypothesis that bats use magnetic fields as a compass calibrator and their primary compass is the sun.

Red foxes (*Vulpes vulpes*) may use magnetoreception when predating small rodents. When foxes perform their high-jumps onto small prey like mice and voles, they tend to jump in a North-Eastern compass direction. In addition, successful attacks are "tightly clustered" to the North.

There is also evidence for magnetoreception in large mammals. Resting and grazing cattle as well as roe deer (*Capreolus capreolus*) and red deer (*Cervus elaphus*) tend to align their body axes in the geomagnetic North-South (N-S) direction. Because wind, sunshine, and slope could be excluded as common ubiquitous factors in this study, alignment toward the vector of the magnetic field provided the most likely explanation for the observed behaviour. However, because of the descriptive nature of this study, alternative explanations (e.g., the sun compass) could not be excluded. In a follow-up study, researchers analyzed body orientations of ruminants in localities where the geomagnetic field is disturbed by high-voltage power lines to determine how local variation in magnetic fields may affect orientation behaviour. This was done by using satellite and aerial images of herds of cattle and field observations of grazing roe deer. Body orientation of both species was random on pastures under or near power lines. Moreover, cattle exposed to various magnetic fields directly beneath or in the vicinity of power lines trending in various magnetic directions exhibited distinct patterns of alignment. The disturbing effect of the power lines on body alignment diminished with the distance from the conductors. In 2011 a group of Czech researchers, however, reported their failed attempt to replicate the finding using different Google Earth images.

In Humans

Magnetic bones have been found in the human nose, specifically the sphenoidal/ethmoid sinuses. Beginning in the late 1970s, the group of Robin Baker at the University of Manchester began to conduct experiments that purported to exhibit magnetoception in humans: people were disoriented and then asked about certain directions; their answers were more accurate if there was no magnet attached to their head. Other scientists have maintained they could not reproduce these results. Some other evidence for human magnetoception was put forward in a 2007 study: low-frequency magnetic fields can produce an evoked response in the brains of human subjects.

Magnetoception in humans has also been achieved by magnetic implants and by non-permanently attached artificial sensory “organs”. However, these exercises do little to demonstrate that humans are innately capable of magnetoreception.

Additionally, a magnetosensitive protein, cryptochrome-2, has been found in the human eye. Given the lack of knowledge as to how cryptochrome mediates magnetosensitivity in Drosophila, it is unclear whether the cryptochrome found in humans functions in the same way and can be used for magnetoception.

Issues

The largest issue affecting verification of an animal magnetic sense is that despite more than 40 years of work on magnetoception there has yet to be an identification of a sensory receptor. Given that the entire receptor system could likely fit in a one-millimetre cube and have a magnetic content of less than one ppm, it is difficult to discern the parts of the brain where this information is processed. In various organisms a cryptochrome mediated receptor has been implicated in magnetoception. At the same time a magnetite system has been found to be relevant to magnetosensation in birds. Furthermore, it is possible that both of these mechanisms play a role in magnetic field detection in animals. This dual mechanism theory in birds raises the questions, if such a mechanism is actually responsible for magnetoception, to what degree is each method responsible for stimulus transduction, and how do they lead to a tranducible signal given a magnetic field with the Earth's strength.

The precise purpose of magnetoception in animal navigation is unclear. Some animals appear to use their magnetic sense as a map, compass, or compass calibrator. The compass method allows animals not only to find north, but also to maintain a constant heading in a particular direction. Although the ability to sense direction is important in migratory navigation, many animals also have the ability to sense small fluctuations in earth's magnetic field to compute coordinate maps with a resolution of a few kilometres or better. For example birds such as the homing pigeon are believed to use the magnetite in their beaks to detect magnetic signposts and thus, the magnetic sense they gain from this pathway is a possible map. Yet, it has also been suggested that homing pigeons and other birds use the visually mediated cryptochrome receptor as a compass.

The purpose of magnetoception in birds and other animals may be varied, but has proved difficult to study, and evidence remains weak. Numerous studies use magnetic fields larger than the Earth's field. Studies such as of *Tritonia* have used electrophysiological recordings from only one or two neurons, and many others have been solely correlational.

Chapter 5

Plate Tectonics

Plate tectonics (from the Late Latin *tectonicus*, from the Greek: τεêτïíéêüò "pertaining to building") is a scientific theory that describes the large-scale motion of Earth's lithosphere. This theoretical model builds on the concept of continental drift which was developed during the first few decades of the 20th century. The geoscientific community accepted the theory after the concepts of seafloor spreading were later developed in the late 1950s and early 1960s.

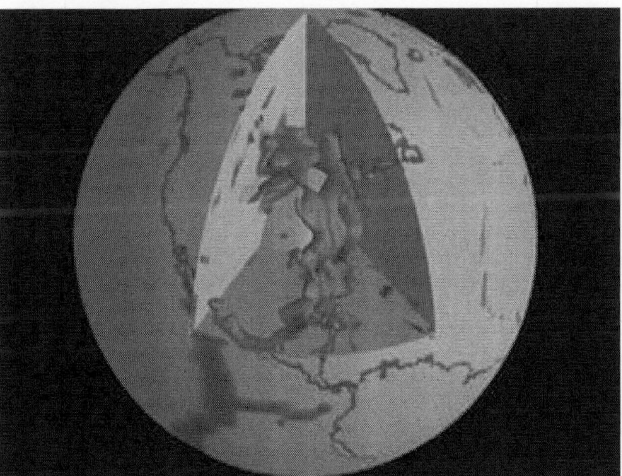

Figure: *Remnants of the Farallon Plate, deep in Earth's mantle. It is thought that much of the plate initially went under North America (particularly the western United States and southwest Canada) at a very shallow angle, creating much of the mountainous terrain in the area (particularly the southern Rocky Mountains).*

The lithosphere, which is the rigid outermost shell of a planet (on Earth, the crust and upper mantle), is broken up into tectonic plates.

On Earth, there are seven or eight major plates (depending on how they are defined) and many minor plates. Where plates meet, their relative motion determines the type of boundary; convergent, divergent, or transform. Earthquakes, volcanic activity, mountain-building, and oceanic trench formation occur along these plate boundaries. The lateral relative movement of the plates typically varies from zero to 100 mm annually.

Tectonic plates are composed of oceanic lithosphere and thicker continental lithosphere, each topped by its own kind of crust. Along convergent boundaries, subduction carries plates into the mantle; the material lost is roughly balanced by the formation of new (oceanic) crust along divergent margins by seafloor spreading. In this way, the total surface of the globe remains the same. This prediction of plate tectonics is also referred to as the conveyor belt principle. Earlier theories (that still have some supporters) propose gradual shrinking (contraction) or gradual expansion of the globe.

Tectonic plates are able to move because the Earth's lithosphere has greater strength than the underlying asthenosphere. Lateral density variations in the mantle result in convection. Plate movement is thought to be driven by a combination of the motion of the seafloor away from the spreading ridge (due to variations in topography and density of the crust, which result in differences in gravitational forces) and drag, with downward suction, at the subduction zones. Another explanation lies in the different forces generated by the rotation of the globe and the tidal forces of the Sun and Moon. The relative importance of each of these factors and their relationship to each other is unclear, and still the subject of much debate.

Key Principles

The outer layers of the Earth are divided into the lithosphere and asthenosphere. This is based on differences in mechanical properties and in the method for the transfer of heat. Mechanically, the lithosphere is cooler and more rigid, while the asthenosphere is hotter and flows more easily. In terms of heat transfer, the lithosphere loses heat by conduction, whereas the asthenosphere also transfers heat by convection and has a nearly adiabatic temperature gradient. This division should not be confused with the *chemical* subdivision of these same layers into the mantle (comprising both the asthenosphere and the mantle portion of the lithosphere) and the crust: a given piece of mantle may be part of the lithosphere or the asthenosphere at different times depending on its temperature and pressure.

The key principle of plate tectonics is that the lithosphere exists as separate and distinct *tectonic plates*, which ride on the fluid-like (visco-elastic solid) asthenosphere. Plate motions range up to a typical 10–40 mm/year (Mid-Atlantic Ridge; about as fast as fingernails grow), to about 160 mm/year (Nazca Plate; about as fast as hair grows). The driving mechanism behind this movement is described below.

Tectonic lithosphere plates consist of lithospheric mantle overlain by either or both of two types of crustal material: oceanic crust (in older texts called *sima* from silicon and magnesium) and continental crust (*sial* from silicon and aluminium). Average oceanic lithosphere is typically 100 km (62 mi) thick; its thickness is a function of its age: as time passes, it conductively cools and subjacent cooling mantle is added to its base. Because it is formed at mid-ocean ridges and spreads outwards, its thickness is therefore a function of its distance from the mid-ocean ridge where it was formed. For a typical distance that oceanic lithosphere must travel before being subducted, the thickness varies from about 6 km (4 mi) thick at mid-ocean ridges to greater than 100 km (62 mi) at subduction zones; for shorter or longer distances, the subduction zone (and therefore also the mean) thickness becomes smaller or larger, respectively. Continental lithosphere is typically ~200 km thick, though this varies considerably between basins, mountain ranges, and stable cratonic interiors of continents. The two types of crust also differ in thickness, with continental crust being considerably thicker than oceanic (35 km vs. 6 km).

The location where two plates meet is called a *plate boundary*. Plate boundaries are commonly associated with geological events such as earthquakes and the creation of topographic features such as mountains, volcanoes, mid-ocean ridges, and oceanic trenches. The majority of the world's active volcanoes occur along plate boundaries, with the Pacific Plate's Ring of Fire being the most active and widely known today. These boundaries are discussed in further detail below. Some volcanoes occur in the interiors of plates, and these have been variously attributed to internal plate deformation and to mantle plumes.

As explained above, tectonic plates may include continental crust or oceanic crust, and most plates contain both. For example, the African Plate includes the continent and parts of the floor of the Atlantic and Indian Oceans. The distinction between oceanic crust and continental crust is based on their modes of formation. Oceanic crust is formed at sea-floor spreading centres, and continental crust is formed through arc volcanism and accretion of terranes through tectonic processes, though some of these terranes may contain ophiolite sequences, which are pieces of oceanic crust considered to be part of the continent when they exit the standard cycle of formation and spreading centres and subduction beneath continents.

Oceanic crust is also denser than continental crust owing to their different compositions. Oceanic crust is denser because it has less silicon and more heavier elements ("mafic") than continental crust ("felsic"). As a result of this density stratification, oceanic crust generally lies below sea level (for example most of the Pacific Plate), while continental crust buoyantly projects above sea level.

Types of Plate Boundaries

Three types of plate boundaries exist, with a fourth, mixed type, characterized by the way the plates move relative to each other. They are associated with different types of surface phenomena. The different types of plate boundaries are:

1. *Transform boundaries (Conservative)* occur where two lithospheric plates slide, or perhaps more accurately, grind past each other along transform faults, where plates are neither created nor destroyed. The relative motion of the two plates is either sinistral (left side toward the observer) or dextral (right side toward the observer). Transform faults occur across a spreading centre. Strong earthquakes can occur along fault. The San Andreas Fault in California is an example of a transform boundary exhibiting dextral motion.
2. *Divergent boundaries (Constructive)* occur where two plates slide apart from each other. At zones of ocean-to-ocean rifting, divergent boundaries form by seafloor spreading, allowing for the formation of new ocean basin. As the continent splits, the ridge forms at the spreading centre, the ocean basin expands, and finally, the plate area increases causing many small volcanoes and/or shallow earthquakes. At zones of continent-to-continent rifting, divergent boundaries may cause new ocean basin to form as the continent splits, spreads, the central rift collapses, and ocean fills the basin. Active zones of Mid-ocean ridges (e.g., Mid-Atlantic Ridge and East Pacific Rise), and continent-to-continent rifting (such as Africa's East African Rift and Valley, Red Sea) are examples of divergent boundaries.
3. *Convergent boundaries (Destructive)* (or *active margins*) occur where two plates slide toward each other to form either a subduction zone (one plate moving underneath the other) or a

continental collision. At zones of continent-to-continent subduction (e.g., Western South America, and Cascade Mountains in Western United States), the dense oceanic lithosphere plunges beneath the less dense continent. Earthquakes then trace the path of the downward-moving plate as it descends into asthenosphere, a trench forms, and as the subducted plate partially melts, magma rises to form continental volcanoes. At zones of ocean-to-ocean subduction (e.g., the Andes mountain range in South America, Aleutian islands, Mariana islands, and the Japanese island arc), older, cooler, denser crust slips beneath less dense crust. This causes earthquakes and a deep trench to form in an arc shape. The upper mantle of the subducted plate then heats and magma rises to form curving chains of volcanic islands. Deep marine trenches are typically associated with subduction zones, and the basins that develop along the active boundary are often called "foreland basins". The subducting slab contains many hydrous minerals which release their water on heating. This water then causes the mantle to melt, producing volcanism. Closure of ocean basins can occur at continent-to-continent boundaries (e.g., Himalayas and Alps): collision between masses of granitic continental lithosphere; neither mass is subducted; plate edges are compressed, folded, uplifted.

4. *Plate boundary zones* occur where the effects of the interactions are unclear, and the boundaries, usually occurring along a broad belt, are not well defined and may show various types of movements in different episodes.

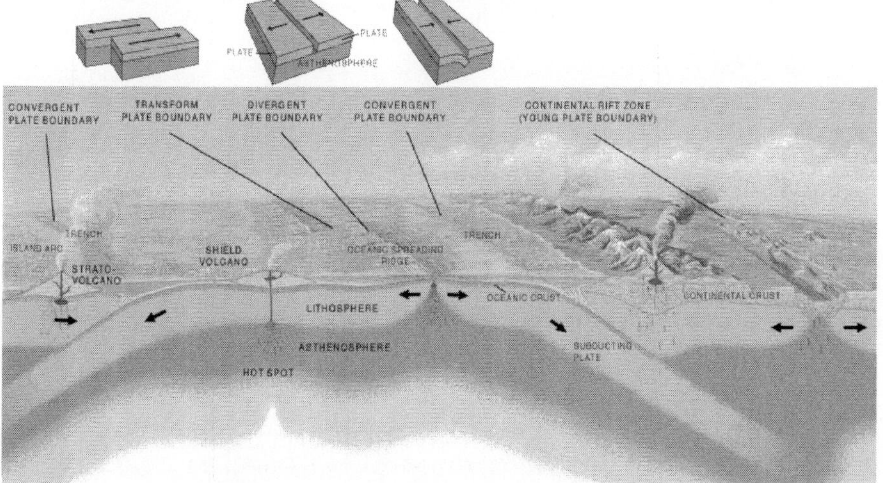

Figure: *Three types of plate boundary.*

Driving Forces of Plate Motion

Plate tectonics is basically a kinematic phenomenon. Scientists agree on the observation and deduction that the plates have moved with respect to one another but continue to debate as to how and when. A major question remains as to what geodynamic mechanism motors plate movement. Here, science diverges in different theories.

It is generally accepted that tectonic plates are able to move because of the relative density of oceanic lithosphere and the relative weakness of the asthenosphere. Dissipation of heat from the mantle is acknowledged to be the original source of the energy required to drive plate tectonics through convection or large scale upwelling and doming. The current view, though still a matter of some debate, asserts that as a consequence, a powerful source of plate motion is generated due to the excess density of the oceanic lithosphere sinking in subduction zones. When the new crust forms at mid-ocean ridges, this oceanic lithosphere is initially less dense than the underlying asthenosphere, but it becomes denser with age as it conductively cools and thickens. The greater density of old lithosphere relative to the underlying asthenosphere allows it to sink into the deep mantle at subduction zones, providing most of the driving force for plate movement.

The weakness of the asthenosphere allows the tectonic plates to move easily towards a subduction zone. Although subduction is believed to be the strongest force driving plate motions, it cannot be the only force since there are plates such as the North American Plate which are moving, yet are nowhere being subducted. The same is true for the enormous Eurasian Plate. The sources of plate motion are a matter of intensive research and discussion among scientists. One of the main points is that the kinematic pattern of the movement itself should be separated clearly from the possible geodynamic mechanism that is invoked as the driving force of the observed movement, as some patterns may be explained by more than one mechanism. In short, the driving forces advocated at the moment can be divided into three categories based on the relationship to the movement: mantle dynamics related, gravity related (mostly secondary forces), and Earth rotation related.

Driving Forces Related to Mantle Dynamics

Mantle Convection: For much of the last quarter century, the leading theory of the driving force behind tectonic plate motions envisaged large scale convection currents in the upper mantle which are transmitted through the asthenosphere. This theory was launched by Arthur Holmes and some forerunners in the 1930s and was immediately recognised as the solution for the acceptance of the theory as originally discussed in the papers of Alfred Wegener in the early years of the century. However, despite its acceptance, it was long debated in the scientific community because the leading ("fixist") theory still envisaged a static Earth without moving continents up until the major break–throughs of the early sixties.

Two– and three–dimensional imaging of Earth's interior (seismic tomography) shows a varying lateral density distribution throughout the mantle. Such density variations can be material (from rock chemistry), mineral (from variations in mineral structures), or thermal (through thermal expansion and contraction from heat energy). The manifestation of this varying lateral density is mantle convection from buoyancy forces.

How mantle convection directly and indirectly relates to plate motion is a matter of ongoing study and discussion in geodynamics. Somehow, this energy must be transferred to the lithosphere for tectonic plates to move. There are essentially two types of forces that are thought to influence plate motion: friction and gravity.

- Basal drag (friction): Plate motion driven by friction between the convection currents in the asthenosphere and the more rigid overlying lithosphere.
- Slab suction (gravity): Plate motion driven by local convection currents that exert a downward pull on plates in subduction zones at ocean trenches. Slab suction may occur in a geodynamic setting where basal tractions continue to act on the plate as it dives into the mantle (although perhaps to a greater extent acting on both the under and upper side of the slab).

Lately, the convection theory has been much debated as modern techniques based on 3D seismic tomography still fail to recognise these predicted large scale convection cells. Therefore, alternative views have been proposed:

In the theory of plume tectonics developed during the 1990s, a modified concept of mantle convection currents is used. It asserts that super plumes rise from the deeper mantle and are the drivers or substitutes of the major convection cells. These ideas, which find their roots in the early 1930s with the so-called "fixistic" ideas of the European and Russian Earth Science Schools, find resonance in the modern theories which envisage hot spots/mantle plumes which remain fixed and are overridden by oceanic and continental lithosphere plates over time and leave their traces in the geological record (though these phenomena are not invoked as real driving mechanisms, but rather as modulators). Modern theories that continue building on the older mantle doming concepts and see plate movements as a secondary phenomena are beyond the scope of this page and are discussed elsewhere (for example on the plume tectonics page).

Another theory is that the mantle flows neither in cells nor large plumes but rather as a series of channels just below the Earth's crust which then provide basal friction to the lithosphere. This theory is called "surge tectonics" and became quite popular in geophysics and geodynamics during the 1980s and 1990s.

Driving Forces Related to Gravity

Forces related to gravity are usually invoked as secondary phenomena within the framework of a more general driving mechanism such as the various forms of mantle dynamics described above.

Gravitational sliding away from a spreading ridge: According to many authors, plate motion is driven by the higher elevation of plates at ocean ridges. As oceanic lithosphere is formed at spreading ridges from hot mantle material, it gradually cools and thickens with age (and thus adds distance from the ridge). Cool oceanic lithosphere is significantly denser than the hot mantle material from which it is derived and so with increasing thickness it gradually subsides into the mantle to compensate the greater load. The result is a slight lateral incline with increased distance from the ridge axis.

This force is regarded as a secondary force and is often referred to as "ridge push". This is a misnomer as nothing is "pushing" horizontally and tensional features are dominant along ridges. It is more accurate to refer to this mechanism as gravitational sliding as variable topography across the totality of the plate can vary considerably and the topography of spreading ridges is only the most prominent feature. Other mechanisms generating this gravitational secondary force include flexural bulging of the lithosphere before it dives underneath an adjacent plate which produces a clear topographical feature that can offset, or at least affect, the influence of topographical ocean ridges, and mantle plumes and hot spots, which are postulated to impinge on the underside of tectonic plates.

Slab-pull: Current scientific opinion is that the asthenosphere is insufficiently competent or rigid to directly cause motion by friction along the base of the lithosphere. Slab pull is therefore most widely thought to be the greatest force acting on the plates. In this current understanding, plate motion is mostly driven by the weight of cold, dense plates sinking into the mantle at trenches. Recent models indicate that trench suction plays an important role as well. However, as the North American Plate is nowhere being subducted, yet it is in motion presents a problem. The same holds for the African, Eurasian, and Antarctic plates.

Gravitational sliding away from mantle doming: According to older theories, one of the driving mechanisms of the plates is the existence of large scale asthenosphere/mantle domes which cause the gravitational sliding of lithosphere plates away from them. This gravitational sliding represents a secondary phenomenon of this basically vertically oriented mechanism. This can act on various scales, from the small scale of one island arc up to the larger scale of an entire ocean basin.

Driving Forces Related to Earth Rotation

Alfred Wegener, being a meteorologist, had proposed tidal forces and pole flight force as the main driving mechanisms behind continental drift; however, these forces were considered far too small to cause continental motion as the concept then was of continents plowing through oceanic crust. Therefore, Wegener later changed his position and asserted that convection currents are the main driving force of plate tectonics in the last edition of his book in 1929.

However, in the plate tectonics context (accepted since the seafloor spreading proposals of Heezen, Hess, Dietz, Morley, Vine, and Matthews during the early 1960s), oceanic crust is suggested to be in motion *with* the continents which caused the proposals related to Earth rotation to be reconsidered. In more recent literature, these driving forces are:

1. Tidal drag due to the gravitational force the Moon (and the Sun) exerts on the crust of the Earth
2. Shear strain of the Earth globe due to N-S compression related to its rotation and modulations;
3. Pole flight force: equatorial drift due to rotation and centrifugal effects: tendency of the plates to move from the poles to the equator ("*Polflucht*");
4. The Coriolis effect acting on plates when they move around the globe;
5. Global deformation of the geoid due to small displacements of rotational pole with respect to the Earth's crust;
6. Other smaller deformation effects of the crust due to wobbles and spin movements of the Earth rotation on a smaller time scale.

For these mechanisms to be overall valid, systematic relationships should exist all over the globe between the orientation and kinematics of deformation and the geographical latitudinal and longitudinal grid of the Earth itself. Ironically, these systematic relations studies in the second half of the nineteenth century and the first half of the twentieth century underline exactly the opposite: that the plates had not moved in time, that the deformation grid was fixed with respect to the Earth equator and axis, and that gravitational driving forces were generally acting vertically and caused only local horizontal movements (the so-called pre-plate tectonic, “fixist theories”). Later studies, therefore, invoked many of the relationships recognised during this pre-plate tectonics period to support their theories.

Of the many forces discussed in this paragraph, tidal force is still highly debated and defended as a possible principle driving force of plate tectonics. The other forces are only used in global geodynamic models not using plate tectonics concepts (therefore beyond the discussions treated in this section) or proposed as minor modulations within the overall plate tectonics model.

In 1973, George W. Moore of the USGS and R. C. Bostrom presented evidence for a general westward drift of the Earth’s lithosphere with respect to the mantle. He concluded that tidal forces (the tidal lag or “friction”) caused by the Earth’s rotation and the forces acting upon it by the Moon are a driving force for plate tectonics. As the Earth spins eastward beneath the moon, the moon’s gravity ever so slightly pulls the Earth’s surface layer back westward, just as proposed by Alfred Wegener.

In a more recent 2006 study, scientists reviewed and advocated these earlier proposed ideas. It has also been suggested recently in Lovett (2006) that this observation may also explain why Venus and Mars have no plate tectonics, as Venus has no moon and Mars’ moons are too small to have significant tidal effects on the planet. In a recent paper, it was suggested that, on the other hand, it can easily be observed that many plates are moving north and eastward, and that the dominantly westward motion of the Pacific ocean basins derives simply from the eastward bias of the Pacific spreading centre (which is not a predicted manifestation of such lunar forces). In the same paper the authors admit, however, that relative to the lower mantle, there is a slight westward component in the motions of all the plates. They demonstrated though that the westward drift, seen only for the past 30 Ma, is attributed to the increased dominance of the steadily growing and accelerating Pacific plate. The debate is still open.

Relative Significance of Each Driving Force Mechanism

The actual vector of a plate's motion is a function of all the forces acting on the plate; however, therein lies the problem regarding what degree each process contributes to the overall motion of each tectonic plate.

The diversity of geodynamic settings and the properties of each plate must clearly result from differences in the degree to which multiple processes are actively driving each individual plate. One method of dealing with this problem is to consider the relative rate at which each plate is moving and to consider the available evidence of each driving force on the plate as far as possible.

One of the most significant correlations found is that lithospheric plates attached to downgoing (subducting) plates move much faster than plates not attached to subducting plates. The Pacific plate, for instance, is essentially surrounded by zones of subduction (the so-called Ring of Fire) and moves much faster than the plates of the Atlantic basin, which are attached (perhaps one could say 'welded') to adjacent continents instead of subducting plates. It is thus thought that forces associated with the downgoing plate (slab pull and slab suction) are the driving forces which determine the motion of plates, except for those plates which are not being subducted. The driving forces of plate motion continue to be active subjects of on-going research within geophysics and tectonophysics.

Development of the Theory

In line with other previous and contemporaneous proposals, in 1912 the meteorologist Alfred Wegener amply described what he called continental drift, expanded in his 1915 book *The Origin of Continents and Oceans* and the scientific debate started that would end up fifty years later in the theory of plate tectonics. Starting from the idea (also expressed by his forerunners) that the present continents once formed a single land mass (which was called Pangea later on) that drifted apart, thus releasing the continents from the Earth's mantle and likening them to "icebergs" of low density granite floating on a sea of denser basalt. Supporting evidence for the idea came from the dove-tailing outlines of South America's east coast and Africa's west coast, and from the matching of the rock formations along these edges. Confirmation of their previous contiguous nature also came from the fossil plants *Glossopteris* and *Gangamopteris*, and the therapsid or mammal-like reptile *Lystrosaurus*, all widely distributed over South America, Africa, Antarctica, India and Australia. The evidence for such an erstwhile joining of these continents was patent to field geologists working in the southern hemisphere. The South African Alex du Toit put together a mass of such information in his 1937 publication *Our Wandering Continents*, and went further than Wegener in recognising the strong links between the Gondwana fragments.

But without detailed evidence and a force sufficient to drive the movement, the theory was not generally accepted: the Earth might have a solid crust and mantle and a liquid core, but there seemed to be no way that portions of the crust could move around. Distinguished scientists, such as Harold Jeffreys and Charles Schuchert, were outspoken critics of continental drift.

Despite much opposition, the view of continental drift gained support and a lively debate started between "drifters" or "mobilists" (proponents of the theory) and "fixists" (opponents). During the 1920s, 1930s and 1940s, the former reached important milestones proposing that convection currents might have driven the plate movements, and that spreading may have occurred below the sea within the oceanic crust. Concepts close to the elements now incorporated in plate tectonics were proposed by geophysicists and geologists (both fixists and mobilists) like Vening-Meinesz, Holmes and Umbgrove.

One of the first pieces of geophysical evidence that was used to support the movement of lithospheric plates came from paleomagnetism. This is based on the fact that rocks of different ages show a variable magnetic field direction, evidenced by studies since the mid–nineteenth century. The magnetic north and south poles reverse through time, and, especially important in paleotectonic studies, the relative position of the magnetic north pole varies through time. Initially, during the first half of the twentieth century, the latter phenomenon was explained by introducing what was called "polar wander", *i.e.*, it was assumed that the north pole location had been shifting through time. An alternative explanation, though, was that the continents had moved (shifted and rotated) relative to the north pole, and each continent, in fact, shows its own "polar wander path". During the late 1950s it was successfully shown on two occasions that these data could show the validity of continental drift: by Keith Runcorn in a paper in 1956, and by Warren Carey in a symposium held in March 1956.

The second piece of evidence in support of continental drift came during the late 1950s and early 60s from data on the bathymetry of the deep ocean floors and the nature of the oceanic crust such as magnetic properties and, more generally, with the development of marine geology which gave evidence for the association of seafloor spreading along the mid-oceanic ridges and magnetic field reversals, published between 1959 and 1963 by Heezen, Dietz, Hess, Mason, Vine & Matthews and Morley.

Simultaneous advances in early seismic imaging techniques in and around Wadati-Benioff zones along the trenches bounding many continental margins, together with many other geophysical (e.g. gravimetric) and geological observations, showed how the oceanic crust could disappear into the mantle, providing the mechanism to balance the extension of the ocean basins with shortening along its margins.

All this evidence, both from the ocean floor and from the continental margins, made it clear around 1965 that continental drift was feasible and the theory of plate tectonics, which was defined in a series of papers between 1965 and 1967, was born, with all its extraordinary explanatory and predictive power. The theory revolutionized the Earth sciences, explaining a diverse range of geological phenomena and their implications in other studies such as paleogeography and paleobiology.

Continental Drift

Continental drift is the movement of the Earth's continents relative to each other, thus appearing to drift across the ocean bed. The speculation that continents might have 'drifted' was first put forward by Abraham Ortelius in 1596. The concept was independently and more fully developed by Alfred Wegener in 1912, but his theory was rejected by some for lack of a mechanism (though this was supplied later by Holmes) and others because of prior theoretical commitments. The idea of continental drift has been subsumed by the theory of plate tectonics, which explains how the continents move.

In the late 19th and early 20th centuries, geologists assumed that the Earth's major features were fixed, and that most geologic features such as basin development and mountain ranges could be explained by vertical crustal movement, described in what is called the geosynclinal theory. Generally, this was placed in the context of a contracting planet Earth due to heat loss in the course of a relatively short geological time. It was observed as early as 1596 that the opposite coasts of the Atlantic Ocean—or, more precisely, the edges of the continental shelves—have similar shapes and seem to have once fitted together.

Since that time many theories were proposed to explain this apparent complementarity, but the assumption of a solid Earth made these various proposals difficult to accept.

Figure: *Alfred Wegener in Greenland in the winter of 1912-13.*

The discovery of radioactivity and its associated heating properties in 1895 prompted a re-examination of the apparent age of the Earth. This had previously been estimated by its cooling rate and assumption the Earth's surface radiated like a black body. Those calculations had implied that, even if it started at red heat, the Earth would have dropped to its present temperature in a few tens of millions of years. Armed with the knowledge of a new heat source, scientists realised that the Earth would be much older, and that its core was still sufficiently hot to be liquid.

By 1915, after having published a first article in 1912, Alfred Wegener was making serious arguments for the idea of continental drift in the first edition of *The Origin of Continents and Oceans*. In that book (re-issued in four successive editions up to the final one in 1936), he noted how the east coast of South America and the west coast of Africa looked as if they were once attached. Wegener was not the first to note this (Abraham Ortelius, Antonio Snider-Pellegrini, Eduard Suess, Roberto Mantovani and Frank Bursley Taylor preceded him just to mention a few), but he was the first to marshal significant fossil and paleo-topographical and climatological evidence to support this simple observation (and was supported in this by researchers such as Alex du Toit). Furthermore, when the rock strata of the margins of separate continents are very similar it suggests that these rocks were formed in the same way, implying that they were joined initially. For instance, parts of Scotland and Ireland contain rocks very similar to those found in Newfoundland and New Brunswick. Furthermore, the Caledonian Mountains of Europe and parts of the Appalachian Mountains of North America are very similar in structure and lithology.

However, his ideas were not taken seriously by many geologists, who pointed out that there was no apparent mechanism for continental drift. Specifically, they did not see how continental rock could plow through the much denser rock that makes up oceanic crust. Wegener could not explain the force that drove continental drift, and his vindication did not come until after his death in 1930.

Evidence of Continental 'Drift'

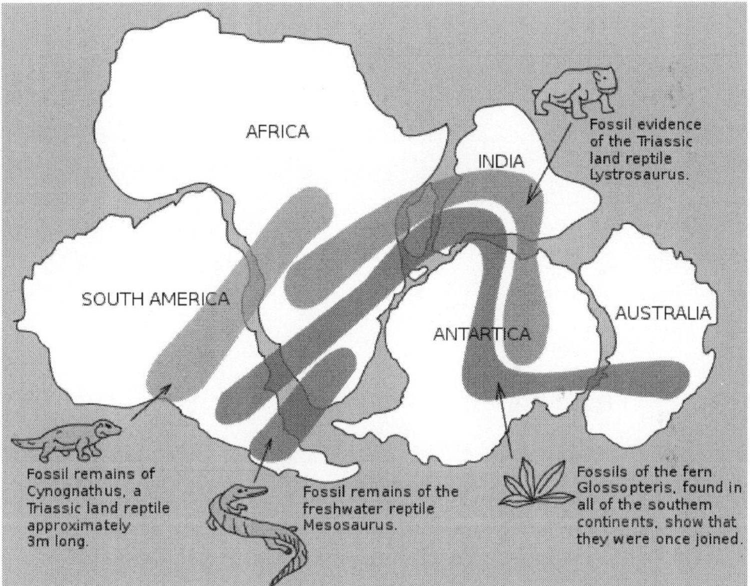

Figure: *Fossil patterns across continents (Gondwanaland).*

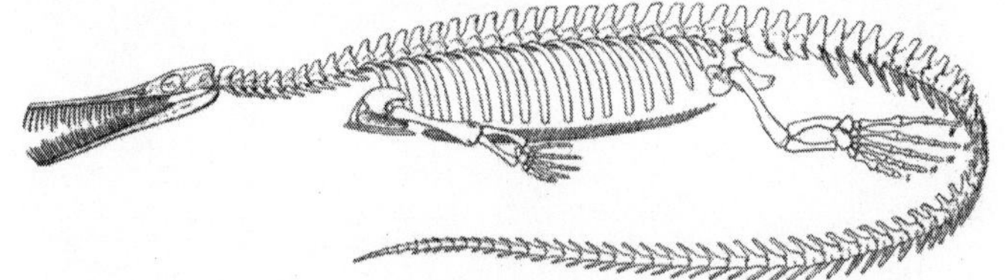

Figure: *Mesosaurus skeleton, MacGregor, 1908.*

Evidence for the movement of continents on tectonic plates is now extensive. Similar plant and animal fossils are found around the shores of different continents, suggesting that they were once joined. The fossils of *Mesosaurus*, a freshwater reptile rather like a small crocodile, found both in Brazil and South Africa, are one example; another is the discovery of fossils of the land reptile *Lystrosaurus* in rocks of the same age at locations in Africa, India, and Antarctica. There is also living evidence—the same animals being found on two continents. Some earthworm families (e.g. Ocnerodrilidae, Acanthodrilidae, Octochaetidae) are found in South America and Africa, for instance.

The complementary arrangement of the facing sides of South America and Africa is obvious, but is a temporary coincidence. In millions of years, slab pull and ridge-push, and other forces of tectonophysics, will further separate and rotate those two continents. It was this temporary feature which inspired Wegener to study what he defined as continental drift, although he did not live to see his hypothesis generally accepted.

Widespread distribution of Permo-Carboniferous glacial sediments in South America, Africa, Madagascar, Arabia, India, Antarctica and Australia was one of the major pieces of evidence for the theory of continental drift. The continuity of glaciers, inferred from oriented glacial striations and deposits called tillites, suggested the existence of the supercontinent of Gondwana, which became a central element of the concept of continental drift. Striations indicated glacial flow away from the equator and toward the poles, based on continents' current positions and orientations, and supported the idea that the southern continents had previously been in dramatically different locations, as well as being contiguous with each other.

Floating Continents, Paleomagnetism, and Seismicity Zones

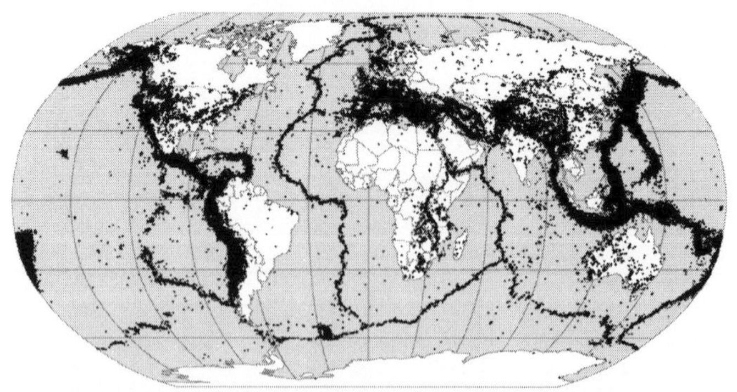

Figure: *Global earthquake epicentres, 1963–1998*

As it was observed early that although granite existed on continents, seafloor seemed to be composed of denser basalt, the prevailing concept during the first half of the twentieth century was that there were two types of crust, named "sial" (continental type crust) and "sima" (oceanic type crust). Furthermore, it was supposed that a static shell of strata was present under the continents. It therefore looked apparent that a layer of basalt (sial) underlies the continental rocks.

However, based on abnormalities in plumb line deflection by the Andes in Peru, Pierre Bouguer had deduced that less-dense mountains must have a downward projection into the denser layer underneath. The concept that mountains had "roots" was confirmed by George B. Airy a hundred years later, during study of Himalayan gravitation, and seismic studies detected corresponding density variations. Therefore, by the mid-1950s, the question remained unresolved as to whether mountain roots were clenched in surrounding basalt or were floating on it like an iceberg.

During the 20th century, improvements in and greater use of seismic instruments such as seismographs enabled scientists to learn that earthquakes tend to be concentrated in specific areas, most notably along the oceanic trenches and spreading ridges. By the late 1920s, seismologists were beginning to identify several prominent earthquake zones parallel to the trenches that typically were inclined 40–60° from the horizontal and extended several hundred kilometres into the Earth. These zones later became known as Wadati-Benioff zones, or simply Benioff zones, in honour of the seismologists who first recognised them, Kiyoo Wadati of Japan and Hugo Benioff of the United States. The study of global seismicity greatly advanced in the 1960s with the establishment of the Worldwide Standardized Seismograph Network (WWSSN) to monitor the compliance of the 1963 treaty banning above-ground testing of nuclear weapons. The much improved data from the WWSSN instruments allowed seismologists to map precisely the zones of earthquake concentration world wide.

Meanwhile, debates developed around the phenomena of polar wander. Since the early debates of continental drift, scientists had discussed and used evidence that polar drift had occurred because continents seemed to have moved through different climatic zones during the past. Furthermore, paleomagnetic data had shown that the magnetic pole had also shifted during time. Reasoning in an opposite way, the continents might have shifted and rotated, while the pole remained relatively fixed. The first time the evidence of magnetic polar wander was used to support the movements of continents was in a paper by Keith Runcorn in 1956, and successive papers by him and his students Ted Irving (who was actually the first to be convinced of the fact that paleomagnetism supported continental drift) and Ken Creer. This was immediately followed by a symposium in Tasmania in March 1956. In this symposium, the evidence was used in the theory of an expansion of the global crust. In this hypothesis the shifting of the continents can be simply explained by a large increase in size of the Earth since its formation. However, this was unsatisfactory because its supporters could offer no convincing mechanism to produce a significant expansion of the Earth. Certainly there is no evidence that the moon has expanded in the past 3 billion years; other work would soon show that the evidence was equally in support of continental drift on a globe with a stable radius.

During the thirties up to the late fifties, works by Vening-Meinesz, Holmes, Umbgrove, and numerous others outlined concepts that were close or nearly identical to modern plate tectonics theory. In particular, the English geologist Arthur Holmes proposed in 1920 that plate junctions might lie beneath the sea, and in 1928 that convection currents within the mantle might be the driving force. Often, these contributions are forgotten because:

- At the time, continental drift was not accepted.
- Some of these ideas were discussed in the context of abandoned fixistic ideas of a deforming globe without continental drift or an expanding Earth.
- They were published during an episode of extreme political and economic instability that hampered scientific communication.
- Many were published by European scientists and at first not mentioned or given little credit in the papers on sea floor spreading published by the American researchers in the 1960s.

Mid-Oceanic Ridge Spreading and Convection

In 1947, a team of scientists led by Maurice Ewing utilizing the Woods Hole Oceanographic Institution's research vessel *Atlantis* and an array of instruments, confirmed the existence of a rise in the central Atlantic Ocean, and found that the floor of the seabed beneath the layer of sediments consisted of basalt, not the granite which is the main constituent of continents. They also found that the oceanic crust was much thinner than continental crust. All these new findings raised important and intriguing questions.

The new data that had been collected on the ocean basins also showed particular characteristics regarding the bathymetry. One of the major outcomes of these datasets was that all along the globe, a system of mid-oceanic ridges was detected. An important conclusion was that along this system, new ocean floor was being created, which led to the concept of the "Great Global Rift". This was described in the crucial paper of Bruce Heezen (1960), which would trigger a real revolution in thinking. A profound consequence of seafloor spreading is that new crust was, and still is, being continually created along the oceanic ridges. Therefore, Heezen advocated the so-called "expanding Earth" hypothesis of S. Warren Carey. So, still the question remained: how can new crust be continuously added along the oceanic ridges without increasing the size of the Earth? In reality, this question had been solved already by numerous scientists during the forties and the fifties, like Arthur Holmes, Vening-Meinesz, Coates and many others: The crust in excess disappeared along what were called the oceanic trenches, where so-called "subduction" occurred. Therefore, when various scientists during the early sixties started to reason on the data at their disposal regarding the ocean floor, the pieces of the theory quickly fell into place.

The question particularly intrigued Harry Hammond Hess, a Princeton University geologist and a Naval Reserve Rear Admiral, and Robert S. Dietz, a scientist with the U.S. Coast and Geodetic Survey who first coined the term *seafloor spreading.* Dietz and Hess (the former published the same idea one year earlier in *Nature,* but priority belongs to Hess who had already distributed an unpublished manuscript of his 1962 article by 1960) were among the small handful who really understood the broad implications of sea floor spreading and how it would eventually agree with the, at that time, unconventional and unaccepted ideas of continental drift and the elegant and mobilistic models proposed by previous workers like Holmes.

In the same year, Robert R. Coats of the U.S. Geological Survey described the main features of island arc subduction in the Aleutian Islands. His paper, though little noted (and even ridiculed) at the time, has since been called "seminal" and "prescient". In reality, it actually shows that the work by the European scientists on island arcs and mountain belts performed and published during the 1930s up until the 1950s was applied and appreciated also in the United States.

If the Earth's crust was expanding along the oceanic ridges, Hess and Dietz reasoned like Holmes and others before them, it must be shrinking elsewhere. Hess followed Heezen, suggesting that new oceanic crust continuously spreads away from the ridges in a conveyor belt–like motion. And, using the mobilistic concepts developed before, he correctly concluded that many millions of years later, the oceanic crust eventually descends along the continental margins where oceanic trenches – very deep, narrow canyons – are formed, e.g. along the rim of the Pacific Ocean basin. The important step Hess made was that convection currents would be the driving force in this process, arriving at the same conclusions as Holmes had decades before with the only difference that the thinning of the ocean crust was performed using Heezen's mechanism of spreading along the ridges. Hess therefore concluded that the Atlantic Ocean was expanding while the Pacific Ocean was shrinking. As old oceanic crust is "consumed" in the trenches (like Holmes and others, he thought this was done by thickening of the continental lithosphere, not, as now understood, by underthrusting at a larger scale of the oceanic crust itself into the mantle), new magma rises and erupts along the spreading ridges to form new crust. In effect, the ocean basins are perpetually being "recycled," with the creation of new crust and the destruction of old oceanic lithosphere occurring simultaneously. Thus, the new mobilistic concepts neatly explained why the Earth does not get bigger with sea floor spreading, why there is so little sediment accumulation on the ocean floor, and why oceanic rocks are much younger than continental rocks.

Seafloor Spreading

Seafloor spreading is a process that occurs at mid-ocean ridges, where new oceanic crust is formed through volcanic activity and then gradually moves away from the ridge. Seafloor spreading helps explain continental drift in the theory of plate tectonics. When oceanic plates diverge, tensional stress causes fractures to occur in the lithosphere. Basaltic magma rises up the fractures and cools on the ocean floor to form new sea floor. Older rocks will be found farther away from the spreading zone while younger rocks will be found nearer to the spreading zone.

Earlier theories (e.g. by Alfred Wegener and Alexander du Toit) of continental drift were that continents "ploughed" through the sea. The idea that the seafloor itself moves (and carries the continents with it) as it expands from a central axis was proposed by Harry Hess from Princeton University in the 1960s. The theory is well accepted now, and the phenomenon is known to be caused by convection currents in the plastic, very weak upper mantle, or asthenosphere.

Incipient Spreading

In the general case, sea floor spreading starts as a rift in a continental land mass, similar to the Red Sea-East Africa Rift System today. The process starts with heating at the base of the continental crust which causes it to become more plastic and less dense. Because less dense objects rise in relation to denser objects, the area being heated becomes a broad dome. As the crust bows upward, fractures occur that gradually grow into rifts. The typical rift system consists of three rift arms at approximately 120 degree angles. These areas are named triple junctions and can be found in several places across the world today. The separated margins of the continents evolve to form passive margins. Hess' theory was that new seafloor is formed when magma is forced upward toward the surface at a mid-ocean ridge.

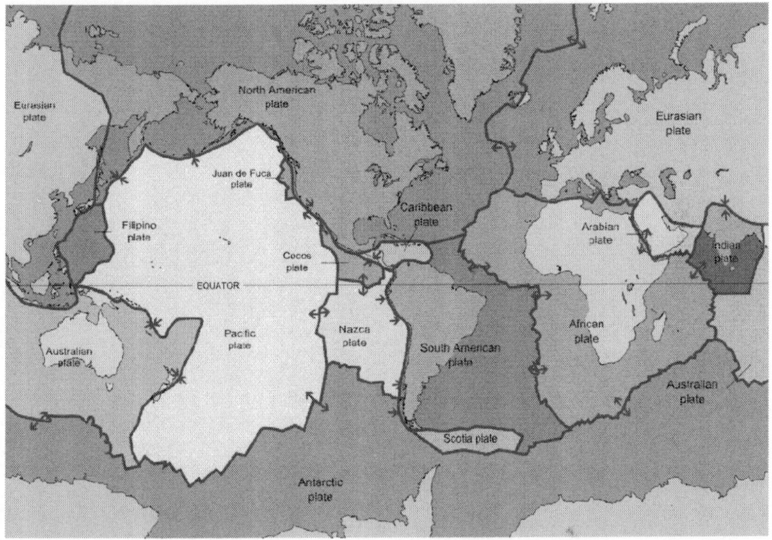

Figure: *Plates in the crust of the earth, according to the plate tectonics theory*

If spreading continues past the incipient stage described above, two of the rift arms will open while the third arm stops opening and becomes a 'failed rift'. As the two active rifts continue to open, eventually the continental crust is attenuated as far as it will stretch. At this point basaltic oceanic crust begins to form between the separating continental fragments. When one of the rifts opens into the existing ocean, the rift system is flooded with seawater and becomes a new sea. The Red Sea is an example of a new arm of the sea. The East African rift was thought to be a "failed" arm that was opening somewhat more slowly than the other two arms, but in 2005 the Ethiopian Afar Geophysical Lithospheric Experiment reported that in the Afar region last September, a 60 km fissure opened as wide as eight metres. During this period of initial flooding the new sea is sensitive to changes in climate and eustasy. As a result the new sea will evaporate (partially or completely) several times before the elevation of the rift valley has been lowered to the point that the sea becomes stable. During this period of evaporation large evaporite deposits will be made in the rift valley. Later these deposits have the potential to become hydrocarbon seals and are of particular interest to petroleum geologists.

Sea floor spreading can stop during the process, but if it continues to the point that the continent is completely severed, then a new ocean basin is created. The Red Sea has not yet completely split Arabia from Africa, but a similar feature can be found on the other side of Africa that has broken completely free. South America once fit into the area of the Niger Delta. The Niger River has formed in the failed rift arm of the triple junction.

Continued Spreading and Subduction

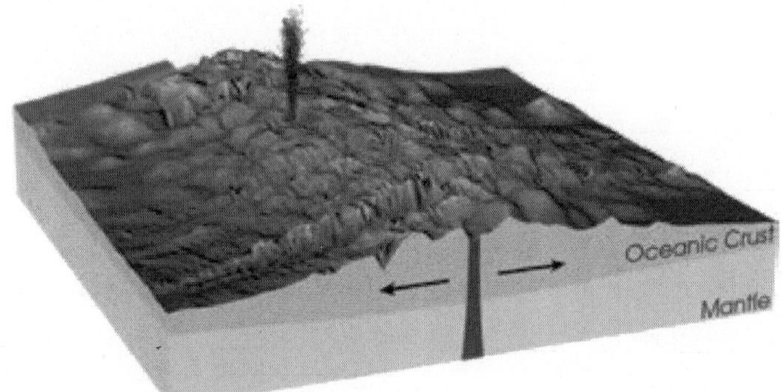

Figure: *Spreading at a mid-ocean ridge*

The new oceanic crust is quite hot relative to old oceanic crust, so the new oceanic basin is shallower than older oceanic basins. If the diameter of the earth remains relatively constant despite the production of new crust, a mechanism must exist by which crust is also destroyed. The destruction of oceanic crust occurs at subduction zones where oceanic crust is forced under either continental crust or oceanic crust. Today, the Atlantic basin is actively spreading at the Mid-Atlantic Ridge. Only a small portion of the oceanic crust produced in the Atlantic is subducted. However, the plates making up the Pacific Ocean are experiencing subduction along many of their boundaries which causes the volcanic activity in what has been termed the Ring of Fire of the Pacific Ocean. The Pacific is also home to one of the world's most active spreading centres (the East Pacific Rise (EPR)) with spreading rates of up to 13 cm/yr. The Mid-Atlantic Ridge is a "textbook" slow-spreading centre, while the EPR is used as an example of fast spreading. The differences in spreading rates affect not only the geometries of the ridges but also the geochemistry of the basalts that are produced.

Since the new oceanic basins are shallower than the old oceanic basins, the total capacity of the world's ocean basins decreases during times of active sea floor spreading. During the opening of the Atlantic Ocean, sea level was so high that a Western Interior Seaway formed across North America from the Gulf of Mexico to the Arctic Ocean.

Debate and Search for Mechanism

At the Mid-Atlantic Ridge (and other places), material from the upper mantle rises through the faults between oceanic plates to form new crust as the plates move away from each other, a phenomenon first observed as continental drift. When Alfred Wegener first presented a hypothesis of continental drift in 1912, he suggested that continents ploughed through the ocean crust. This was impossible: oceanic crust is both more dense and more rigid than continental crust. Accordingly Wegener's theory wasn't taken very seriously, especially in the United States.

Since then, it has been shown that the motion of the continents is linked to seafloor spreading. In the 1960s, the past record of geomagnetic reversals was noticed by observing the magnetic stripe "anomalies" on the ocean floor. This results in broadly evident "stripes" from which the past magnetic field polarity can be inferred by looking at the data gathered from simply towing a magnetometre on the sea surface or from an aircraft. The stripes on one side of the mid-ocean ridge were the mirror image of those on the other side. The seafloor must have originated on the Earth's great fiery welts, like the Mid-Atlantic Ridge and the East Pacific Rise.

The driver for seafloor spreading in plates with active margins is the weight of the cool, dense, subducting slabs that pull them along. The magmatism at the ridge is considered to be "passive upswelling", which is caused by the plates being pulled apart under the weight of their own slabs. This can be thought of as analogous to a rug on a table with little friction: when part of the rug is off of the table, its weight pulls the rest of the rug down with it.

Sea Floor Global Topography: Half-Space Model

To first approximation, sea floor global topography in areas without significant subduction can be estimated by the half-space model. In this model, the seabed height is determined by the oceanic lithosphere temperature, due to thermal expansion. Oceanic lithosphere is continuously formed at a constant rate at the mid-ocean ridges. The source of the lithosphere has a half-plane shape ($x = 0$, $z < 0$) and a constant temperature T_1. Due to its continuous creation, the lithosphere at $x > 0$ is moving away from the ridge at a constant velocity v, which is assumed large compared to other typical scales in the problem. The temperature at the upper boundary of the lithosphere (z=0) is a constant $T_0 = 0$. Thus at x = 0 the temperature is the Heaviside step function $T_1 \cdot \Theta(-z)$. Finally, we assume the system is at a quasi-steady state, so that the temperature distribution is constant in time, i.e. T=T(x,z).

By calculating in the frame of reference of the moving lithosphere (velocity v), which have spatial coordinate x' = x-vt, we may write T = T(x',z,t) and use the heat equation: $\frac{\partial T}{\partial t} = \kappa \nabla^2 T = \kappa \frac{\partial^2 T}{\partial^2 z} + \kappa \frac{\partial^2 T}{\partial^2 x'}$ where κ is the thermal diffusivity of the mantle lithosphere.

Since T depends on x' and t only through the combination $x = x' + vt$, we have: $\frac{\partial T}{\partial x'} = \frac{1}{v} \cdot \frac{\partial T}{\partial t}$

Thus: $\frac{\partial T}{\partial t} = \kappa \nabla^2 T = \kappa \frac{\partial^2 T}{\partial^2 z} + \frac{\kappa}{v^2} \frac{\partial^2 T}{\partial^2 t}$

We now use the assumption that v is large compared to other scales in the problem; we therefore neglect the last term in the equation, and get a 1-dimensional diffusion equation: $\frac{\partial T}{\partial t} = \kappa \frac{\partial^2 T}{\partial^2 z}$ with the initial conditions $T(t = 0) = T_1 \cdot \Theta(-z)$.

The solution for $z \le 0$ is given by the error function erf :

$$T(x', z, t) = T_1 \cdot \mathrm{erf}\left(\frac{z}{2\sqrt{\kappa t}}\right).$$

Due to the large velocity, the temperature dependence on the horizontal direction is negligible, and the height at time t (i.e. of sea floor of age t) can be calculated by integrating the thermal expansion over z:

where α_{eff} is the effective volumetric thermal expansion coefficient, and h_0 is the mid-ocean ridge height (compared to some reference).

Note that the assumption the v is relatively large is equivalently to the assumption that the thermal diffusivity κ is small compared to L^2 / T, where L is the acean width (from mid-ocean ridges to continental shelf) and T is its age.

The effective thermal expansion coefficient α_{eff} is different from the usual thermal expansion coefficient α due to isostasic effect of the change in water column height above the lithosphere as it expands or retracts. Both coefficients are related by:

$$\alpha_{eff} = \alpha \cdot \frac{\rho}{\rho - \rho_w}$$

where $\rho \sim 3.3g/cm^3$ is the rock density and $\rho_0 = 1g/cm^3$ is the density of water.

By substituting the parameters by their rough estimates: $\kappa \sim 8{\cdot}10^{-7}$ m²/sec, $\alpha \sim 4{\cdot}10^{-5}$ °C⁻¹ and T_1 ~1220 °C (for the Atlantic and Indian oceans) or ~1120 °C (for the eastern Pacific), we have:

$$h(t) \sim h_0 - 350\sqrt{t}$$

for the eastern Pacific Ocean, and:

$$h(t) \sim h_0 - 390\sqrt{t}$$

for the Atlantic and Indian Ocean, where the height is in metres and time is in millions of years. To get the dependence on x, one must substitute t = x/v ~ Tx/L, where L is the distance between the ridge to the continental shelf (roughly half the ocean width), and T is the ocean age.

Magnetic Striping

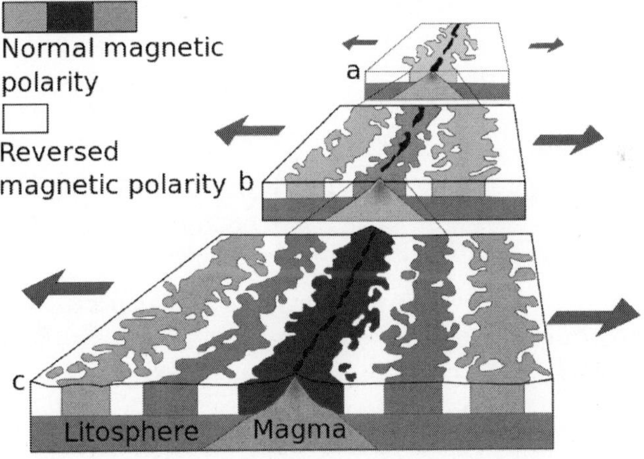

Figure: *Seafloor magnetic striping.*

Beginning in the 1950s, scientists like Victor Vacquier, using magnetic instruments (magnetometres) adapted from airborne devices developed during World War II to detect submarines, began recognising odd magnetic variations across the ocean floor. This finding, though unexpected, was not entirely surprising because it was known that basalt—the iron-rich, volcanic rock making up the ocean floor—contains a strongly magnetic mineral (magnetite) and can locally distort compass readings. This distortion was recognised by Icelandic mariners as early as the late 18th century. More important, because the presence of magnetite gives the basalt measurable magnetic properties, these newly discovered magnetic variations provided another means to study the deep ocean floor. When newly formed rock cools, such magnetic materials recorded the Earth's magnetic field at the time.

As more and more of the seafloor was mapped during the 1950s, the magnetic variations turned out not to be random or isolated occurrences, but instead revealed recognisable patterns. When these magnetic patterns were mapped over a wide region, the ocean floor showed a zebra-like pattern: one stripe with normal polarity and the adjoining stripe with reversed polarity. The overall pattern, defined by these alternating bands of normally and reversely polarized rock, became known as magnetic striping, and was published by Ron G. Mason and co-workers in 1961, who did not find, though, an explanation for these data in terms of sea floor spreading, like Vine, Matthews and Morley a few years later.

The discovery of magnetic striping called for an explanation. In the early 1960s scientists such as Heezen, Hess and Dietz had begun to theorise that mid-ocean ridges mark structurally weak zones where the ocean floor was being ripped in two lengthwise along the ridge crest. New magma from deep within the Earth rises easily through these weak zones and eventually erupts along the crest of the ridges to create new oceanic crust. This process, at first denominated the "conveyer belt hypothesis" and later called seafloor spreading, operating over many millions of years continues to form new ocean floor all across the 50,000 km-long system of mid–ocean ridges.

Only four years after the maps with the "zebra pattern" of magnetic stripes were published, the link between sea floor spreading and these patterns was correctly placed, independently by Lawrence Morley, and by Fred Vine and Drummond Matthews, in 1963 now called the Vine-Matthews-Morley hypothesis. This hypothesis linked these patterns to geomagnetic reversals and was supported by several lines of evidence:

1. the stripes are symmetrical around the crests of the mid-ocean ridges; at or near the crest of the ridge, the rocks are very young, and they become progressively older away from the ridge crest;
2. the youngest rocks at the ridge crest always have present-day (normal) polarity;
3. stripes of rock parallel to the ridge crest alternate in magnetic polarity (normal-reversed-normal, etc.), suggesting that they were formed during different epochs documenting the (already

known from independent studies) normal and reversal episodes of the Earth's magnetic field.

By explaining both the zebra-like magnetic striping and the construction of the mid-ocean ridge system, the seafloor spreading hypothesis (SFS) quickly gained converts and represented another major advance in the development of the plate-tectonics theory. Furthermore, the oceanic crust now came to be appreciated as a natural "tape recording" of the history of the geomagnetic field reversals (GMFR) of the Earth's magnetic field. Today, extensive studies are dedicated to the calibration of the normal-reversal patterns in the oceanic crust on one hand and known timescales derived from the dating of basalt layers in sedimentary sequences (magnetostratigraphy) on the other, to arrive at estimates of past spreading rates and plate reconstructions.

Definition and Refining of the Theory

After all these considerations, Plate Tectonics (or, as it was initially called "New Global Tectonics") became quickly accepted in the scientific world, and numerous papers followed that defined the concepts:

- In 1965, Tuzo Wilson who had been a promotor of the sea floor spreading hypothesis and continental drift from the very beginning added the concept of transform faults to the model, completing the classes of fault types necessary to make the mobility of the plates on the globe work out.
- A symposium on continental drift was held at the Royal Society of London in 1965 which must be regarded as the official start of the acceptance of plate tectonics by the scientific community, and which abstracts are issued as Blacket, Bullard & Runcorn (1965). In this symposium, Edward Bullard and co-workers showed with a computer calculation how the continents along both sides of the Atlantic would best fit to close the ocean, which became known as the famous "Bullard's Fit".
- In 1966 Wilson published the paper that referred to previous plate tectonic reconstructions, introducing the concept of what is now known as the "Wilson Cycle".
- In 1967, at the American Geophysical Union's meeting, W. Jason Morgan proposed that the Earth's surface consists of 12 rigid plates that move relative to each other.
- Two months later, Xavier Le Pichon published a complete model based on 6 major plates with their relative motions, which marked the final acceptance by the scientific community of plate tectonics.

- In the same year, McKenzie and Parker independently presented a model similar to Morgan's using translations and rotations on a sphere to define the plate motions.

Implications for Biogeography

Continental drift theory helps biogeographers to explain the disjunct biogeographic distribution of present day life found on different continents but having similar ancestors. In particular, it explains the Gondwanan distribution of ratites and the Antarctic flora.

Plate Reconstruction

Plate reconstruction is the process of reconstructing the positions of tectonic plates relative to each other (relative motion) or to other reference frames, such as the earth's magnetic field or groups of hotspots, in the geological past. This helps determine the shape and make-up of ancient supercontinents and provides a basis for paleogeographic reconstructions.

Reconstruction is used to establish past (and future) plate configurations, helping determine the shape and make-up of ancient supercontinents and providing a basis for paleogeography.

Defining Plate Boundaries

Current plate boundaries are defined by their seismicity. Past plate boundaries within existing plates are identified from a variety of evidence, such as the presence of ophiolites that are indicative of vanished oceans.

Past Plate Motions

Tectonic motion first began around three billion years ago. Various types of quantitative and semi-quantitative information are available to constrain past plate motions. The geometric fit between continents, such as between west Africa and South America is still an important part of plate reconstruction. Magnetic stripe patterns provide a reliable guide to relative plate motions going back into the Jurassic period. The tracks of hotspots give absolute reconstructions, but these are only available back to the Cretaceous. Older reconstructions rely mainly on paleomagnetic pole data, although these only constrain the latitude and rotation, but not the longitude. Combining poles of different ages in a particular plate to produce apparent polar wander paths provides a method for comparing the motions of different plates through time. Additional evidence comes from the distribution of certain sedimentary rock types, faunal provinces shown by particular fossil groups, and the position of orogenic belts.

Formation and Break-up of Continents

The movement of plates has caused the formation and break-up of continents over time, including occasional formation of a supercontinent that contains most or all of the continents. The supercontinent Columbia or Nuna formed during a period of 2,000 to 1,800 million years ago and broke up about 1,500 to 1,300 million years ago.

The supercontinent Rodinia is thought to have formed about 1 billion years ago and to have embodied most or all of Earth's continents, and broken up into eight continents around 600 million years ago. The eight continents later re-assembled into another supercontinent called Pangaea; Pangaea broke up into Laurasia (which became North America and Eurasia) and Gondwana (which became the remaining continents).

The Himalayas, the world's tallest mountain range, are assumed to have been formed by the collision of two major plates. Before uplift, they were covered by the Tethys Ocean.

Current Plates

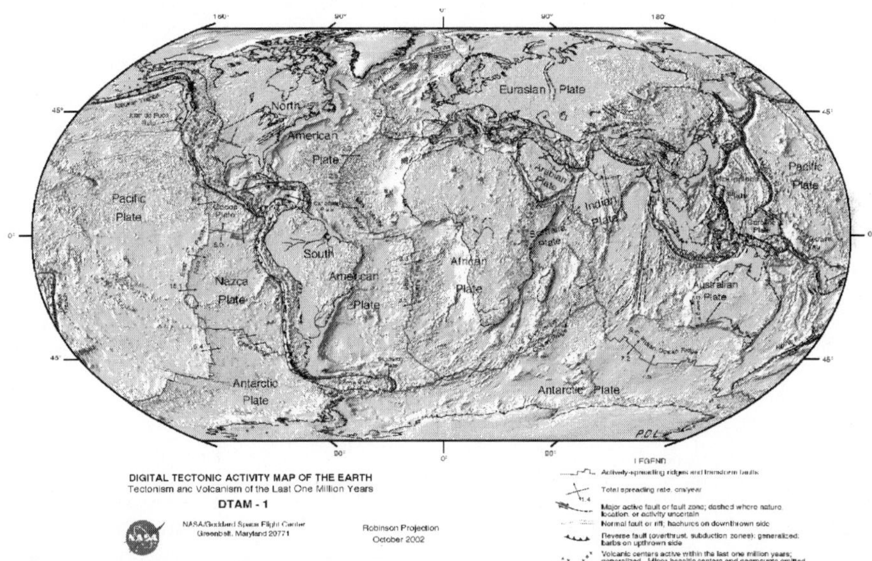

Figure: *Depending on how they are defined, there are usually seven or eight "major" plates: African, Antarctic, Eurasian, North American, South American, Pacific, and Indo-Australian. The latter is sometimes subdivided into the Indian and Australian plates.*

There are dozens of smaller plates, the seven largest of which are the Arabian, Caribbean, Juan de Fuca, Cocos, Nazca, Philippine Sea and Scotia.

The current motion of the tectonic plates is today determined by remote sensing satellite data sets, calibrated with ground station measurements.

Other Celestial Bodies (Planets, Moons)

The appearance of plate tectonics on terrestrial planets is related to planetary mass, with more massive planets than Earth expected to exhibit plate tectonics. Earth may be a borderline case, owing its tectonic activity to abundant water (silica and water form a deep eutectic.)

Venus

Venus shows no evidence of active plate tectonics. There is debatable evidence of active tectonics in the planet's distant past; however, events taking place since then (such as the plausible and generally accepted hypothesis that the Venusian lithosphere has thickened greatly over the course of several hundred million years) has made constraining the course of its geologic record difficult. However, the numerous well-preserved impact craters have been utilized as a dating method to approximately date the Venusian surface (since there are thus far no known samples of Venusian rock to be dated by more reliable methods). Dates derived are dominantly in the range 500 to 750 million years ago, although ages of up to 1,200 million years ago have been calculated. This research has led to the fairly well accepted hypothesis that Venus has undergone an essentially complete volcanic resurfacing at least once in its distant past, with the last event taking place approximately within the range of estimated surface ages. While the mechanism of such an impressive thermal event remains a debated issue in Venusian geosciences, some scientists are advocates of processes involving plate motion to some extent.

One explanation for Venus' lack of plate tectonics is that on Venus temperatures are too high for significant water to be present. The Earth's crust is soaked with water, and water plays an important role in the development of shear zones. Plate tectonics requires weak surfaces in the crust along which crustal slices can move, and it may well be that such weakening never took place on Venus because of the absence of water. However, some researchers remain convinced that plate tectonics is or was once active on this planet.

Mars

Mars is considerably smaller than Earth and Venus, and there is evidence for ice on its surface and in its crust.

In the 1990s, it was proposed that Martian Crustal Dichotomy was created by plate tectonic processes. Scientists today disagree, and believe that it was created either by upwelling within the Martian mantle that thickened the crust of the Southern Highlands and formed Tharsis or by a giant impact that excavated the Northern Lowlands.

Valles Marineris may be a Tectonic Boundary.

Observations made of the magnetic field of Mars by the *Mars Global Surveyor* spacecraft in 1999 showed patterns of magnetic striping discovered on this planet. Some scientists interpreted these as requiring plate tectonic processes, such as seafloor spreading. However, their data fail a "magnetic reversal test", which is used to see if they were formed by flipping polarities of a global magnetic field.

Galilean Satellites of Jupiter

Some of the satellites of Jupiter have features that may be related to plate-tectonic style deformation, although the materials and specific mechanisms may be different from plate-tectonic activity on Earth. On 8 September 2014, NASA reported finding evidence of plate tectonics on Europa, a satellite of Jupiter - the first sign of such geological activity on another world other than Earth.

Titan, Moon of Saturn

Titan, the largest moon of Saturn, was reported to show tectonic activity in images taken by the Huygens Probe, which landed on Titan on January 14, 2005.

Exoplanets

On Earth-sized planets, plate tectonics is more likely if there are oceans of water; however, in 2007, two independent teams of researchers came to opposing conclusions about the likelihood of plate tectonics on larger super-earths with one team saying that plate tectonics would be episodic or stagnant and the other team saying that plate tectonics is very likely on super-earths even if the planet is dry.

Geomagnetic Jerk

In geophysics, a geomagnetic jerk or secular geomagnetic variation impulse is a relatively sudden change in the second derivative of the Earth's magnetic field with respect to time.

These events were noted by Courtillot and others in 1978. The clearest ones, observed all over the world, happened in 1969, 1978, 1991, and 1999. Data before 1969 is scarcer, but there is evidence of other global jerks in 1901, 1913, and 1925. Other events in 1932, 1949, 1958, 1986, and 2003 were detected only in some parts of the world. These events are believed to originate in the interior of the Earth (rather than being due to external phenomena such as the solar wind); but their precise cause is still a matter of research.

The name "jerk" was borrowed from dynamics, where it means the rate of change of the acceleration of a body, that is, the third derivative of its position with respect to time (the acceleration being the second derivative); or, more specifically, a sudden and momentary spike (or dip) in that rate.

Description

Jerks seem to occur in irregular intervals, in the average about once every 10 years. In the period between jerks, each component of the field at a specific location changes with time t approximately as a fixed polynomial of the second degree, $A\,t^2 + B\,t + C$. Each jerk is a relatively sudden change (spread over a period of a few months to a couple of years) in the A coefficient of this formula, which determines the second derivative; and usually in B and C coefficients as well.

The strength of each jerk varies from location to location, and some jerks are observed only in some regions. For example, the 1949 jerk was clearly observed at Tucson (North America, long. 249.17°), but not at Chambon la Forêt (Europe, long. 2.27°). Moreover, the global jerks seem to occur at slightly different times in different regions; often earlier in the Northern hemisphere than in the Southern hemisphere.

Theories

These events are believed to be caused by changes in the flow patterns of the liquid outer core of the Earth.

Another theory is that they are due to torsional oscillations in the solid inner core of the Earth. There have been claims that they are connected to strong earthquakes.

Curie's Law

In a paramagnetic material the magnetization of the material is (approximately) directly proportional to an applied magnetic field. However, if the material is heated, this proportionality is reduced: for a fixed value of the field, the magnetization is (approximately) inversely proportional to temperature. This fact is encapsulated by Curie's law:

$$\mathbf{M} = C \cdot \frac{\mathbf{B}}{T},$$

where

M is the resulting magnetisation

B is the magnetic field, measured in teslas

T is absolute temperature, measured in kelvins

C is a material-specific Curie constant.

This relation was discovered experimentally (by fitting the results to a correctly guessed model) by Pierre Curie. It only holds for high temperatures, or weak magnetic fields. As the derivations below show, the magnetization saturates in the opposite limit of low temperatures, or strong fields.

Derivation with Quantum Mechanics

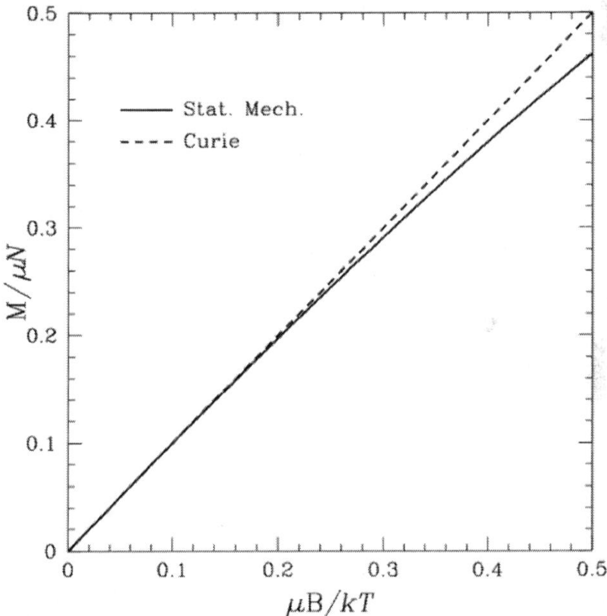

Figure: *Magnetization of a paramagnet as a function of inverse temperature.*

A simple model of a paramagnet concentrates on the particles which compose it which do not interact with each other. Each particle has a magnetic moment given by $\vec{\mu}$. The energy of a magnetic moment in a magnetic field is given by

$$E = -\vec{\mu}\cdot\vec{B}.$$

Two-state (spin-1/2) particles

To simplify the calculation, we are going to work with a 2-state particle: it may either align its magnetic moment with the magnetic field, or against it. So the only possible values of magnetic moment are then μ and $-\mu$. If so, then such a particle has only two possible energies

$$E_0 = -\mu B$$

and

$$E_1 = \mu B.$$

When one seeks the magnetization of a paramagnet, one is interested in the likelihood of a particle to align itself with the field. In other words, one seeks the expectation value of the magnetization μ:

$$\langle \mu \rangle = \mu P(\mu) + (-\mu)P(-\mu) = \frac{1}{Z}\left(\mu e^{\mu B\beta} - \mu e^{-\mu B\beta}\right) = \frac{2\mu}{Z}\sinh(\mu B\beta),$$

where the probability of a configuration is given by its Boltzmann factor, and the partition function Z provides the necessary normalization for probabilities (so that the sum of all of them is unity.) The partition function of one particle is:

$$Z = \sum_{n=0,1} e^{-E_n\beta} = e^{\mu B\beta} + e^{-\mu B\beta} = 2\cosh(\mu B\beta).$$

Therefore, in this simple case we have:

$$\langle \mu \rangle = \mu \tanh(\mu B\beta).$$

This is magnetization of one particle, the total magnetization of the solid is given by

$$M = N\langle \mu \rangle = N\mu \tanh\left(\frac{\mu B}{kT}\right)$$

The formula above is known as the Langevin paramagnetic equation. Pierre Curie found an approximation to this law which applies to the relatively high temperatures and low magnetic fields used in his experiments. Let's see what happens to the magnetization as we specialise it to large T and small B. As temperature increases and magnetic field decreases, the argument of hyperbolic tangent decreases. Another way to say this is

$$\left(\frac{\mu B}{kT}\right) \ll 1$$

this is sometimes called the Curie regime. We also know that if $|x| \ll 1$, then

$$\tanh x \approx x \cdot$$

So,

$$\mathbf{M}(T \to \infty) = \frac{N\mu^2}{k}\frac{\mathbf{B}}{T},$$

with a Curie constant given by $C = N\mu^2 / k$. Also, in the opposite regime of low temperatures or high fields, M tends to a maximum value of $N\mu$, corresponding to all the particles being completely aligned with the field.

General Case

When the particles have an arbitrary spin (any number of spin states), the formula is a bit more complicated. At low magnetic fields or high temperature, the spin follows Curie's law, with

$$C = \frac{\mu_B^2}{3k_B} Ng^2 J(J+1)$$

where J is the total angular momentum quantum number and g is the spin's g-factor (such that $\mu = gJ\mu_B$ is the magnetic moment).

As the spin approaches infinity, the formula for the magnetization approaches the classical value derived in the following section.

Derivation with Classical Statistical Mechanics

An alternative treatment applies when the paramagnetons are imagined to be classical, freely-rotating magnetic moments. In this case, their position will be determined by their angles in spherical coordinates, and the energy for one of them will be:

$$E = -\mu B \cos\theta,$$

where θ is the angle between the magnetic moment and the magnetic field (which we take to be pointing in the z coordinate.) The corresponding partition function is

$$Z = \int_0^{2\pi} d\phi \int_0^{\pi} d\theta \sin\theta \exp(\mu B\beta\cos\theta).$$

We see there is no dependence on the ϕ angle, and also we can change variables to $y = \cos\theta$ to obtain

$$Z = 2\pi\int_{-1}^{1} dy \exp(\mu B\beta y) = 2\pi\frac{\exp(\mu B\beta) - \exp(-\mu B\beta)}{\mu B\beta} = \frac{4\pi\sinh(\mu B\beta)}{\mu B\beta.}$$

Now, the expected value of the z component of the magnetization (the other two are seen to be null (due to integration over ϕ), as they should) will be given by

$$\langle \mu_z \rangle = \frac{1}{Z} \int_0^{2\pi} d\phi \int_0^{\pi} d\theta \sin\theta \exp(\mu B \beta \cos\theta) [\mu \cos\theta].$$

To simplify the calculation, we see this can be written as a differentiation of :

$$\langle \mu_z \rangle = \frac{1}{ZB} \partial_\beta Z.$$

(This approach can also be used for the model above, but the calculation was so simple this is not so helpful.)

Carrying out the derivation we find

$$\langle \mu_z \rangle = \mu L(\mu B \beta),$$

where L is the Langevin function:

$$L(x) = \coth x - \frac{1}{x}.$$

This function would appear to be singular for small x, but it is not, since the two singular terms cancel each other. In fact, its behaviour for small arguments is $L(x) \approx x/3$, so the Curie limit also applies, but with a Curie constant three times smaller in this case. Similarly, the function saturates at 1 for large values of its argument, and the opposite limit is likewise recovered.

Curie-Weiss Law

The Curie–Weiss law describes the magnetic susceptibility ÷ of a ferromagnet in the paramagnetic region above the Curie point:

$$\chi = \frac{C}{T - T_c}$$

where C is a material-specific Curie constant, T is absolute temperature, measured in kelvins, and T_c is the Curie temperature, measured in kelvin. The law predicts a singularity in the susceptibility at $T = T_c$. Below this temperature the ferromagnet has a spontaneous magnetization.

The magnetic moment of a magnet is a quantity that determines the torque and force it will experience in an external magnetic field. A loop of electric current, a bar magnet, an electron, a molecule, and a planet all have magnetic moments.

The magnetization or magnetic polarization is the vector field that expresses the density of permanent or induced magnetic moment in a magnetic material. The origin of the magnetic moments responsible for magnetization can be either microscopic electric current resulting from the motion of electron in atom, or the spin of the electrons or the nuclei. Net magnetization results from the response of a material to an external magnetic field, together with any unbalanced magnetic moment that may be present even in the absence of the external magnetic field; for example, in cold enough iron. We call the latter spontaneous magnetization. Some other materials share this property with iron as well. For example, Nickel and magnetite. These are called ferromagnet. The threshold temperature (depends on the material) below which a material is ferromagnetic is called the Curie temperature.

Limitation of Curie-Weiss law

In many materials the Curie–Weiss law fails to describe the susceptibility in the immediate vicinity of the Curie point, since it is based on a mean-field approximation. Instead, there is a critical behaviour of the form

$$\chi \sim \frac{1}{(T - T_c)^{\gamma}}$$

with the critical exponent γ. However, at temperatures $T \gg T_c$ the expression of the Curie–Weiss law still holds, but with T_c replaced by a temperature Θ that is somewhat higher than the actual Curie temperature. Some authors call Θ the Weiss constant to distinguish it from the temperature of the actual Curie point.

Classical approaches to magnetic susceptibility and Bohr–van Leeuwen theorem According to Bohr–van Leeuwen theorem when statistical mechanics and classical mechanics are applied consistently, the thermal average of the magnetization is always zero. Magnetism cannot be explained without quantum mechanics. However we list some classical approaches to it as they are easy to understand and relate to even though they are incorrect.

The magnetic moment of a free atom is due to the orbital angular momentum and spin of its electrons and nucleus. When the atoms are such that their shells are completely filled they do not have any net magnetic dipole moment in the absence of external magnetic field. When present, such field distorts the trajectories (classical concept) of the electrons so that the applied field could be opposed as predicted by the Lenz's law. In other words the net magnetic dipole induced by the external field is in the opposite direction and such materials are repelled by it. These are called diamagnetic materials.

Sometimes an atom has a net magnetic dipole moment even in the absence of an external magnetic field. The contributions of the individual electrons and nucleus to the total angular momentum do not cancel each other. This happens when the shells of the atoms are not fully filled up (Hund's Rule). A collection of such atoms however may not have any net magnetic moment as these dipoles are not aligned. An external magnetic field may serve to align them to some extent and develop a net magnetic moment per volume. Such alignment is temperature dependent as thermal agitation acts to disorient the dipoles. Such materials are called paramagnetic.

In some materials, the atoms (with net magnetic dipole moments) can interact with each other to align themselves even in the absence of any external magnetic field when the thermal agitation is low enough. Alignment could be parallel (ferromagnetism) or anti-parallel. In case of anti-parallel, the dipole moments may or may not cancel each other (antiferromagnetism, ferrimagnetism).

Density Matrix Approach to Magnetic Susceptibility

We take a very simple situation in which each atom can be approximated as a two state system. The thermal energy is so low that the atom is in ground state. In this ground state the atom is assumed to have no net orbital angular momentum but only one unpaired electron to give it a spin of half. In the presence of an external magnetic field the ground state will split into two states having energy difference proportional to the applied field. The spin of the unpaired electron is parallel to the field in the higher energy state and anti-parallel in the lower one.

A density matrix, ρ, is a matrix that describes a quantum system in a mixed state, a statistical ensemble of several quantum states (here several similar 2-state atoms). This should be contrasted with a single state vector that describes a quantum system in a pure state. The expectation value of a measurement, A, over the ensemble is $\langle A \rangle = Tr(A\rho)$. In terms of a complete set of states, $| i \rangle$, one can write

$$\rho = \sum_{ij} \rho_{ij} \, | i \rangle \langle j | .$$

Von Neumann's equation tells us how the density matrix evolves with time.

$$i\hbar \frac{d}{dt} \rho(t) = [H, \rho(t)]$$

In equilibrium, one has $[H, \rho] = 0$, and the allowed density matrices are $f(H)$. The canonical ensemble has $\rho = \exp(-H / T) / Z$ where $Z = Tr\exp(-H / T)$.

For the 2-state system, we can write $H = -\gamma\hbar B\sigma_3$. Here γ is the gyromagnetic ratio. Hence $Z = 2\cosh(\gamma\hbar B / (2T))$, and

$$\rho(B,T) = \frac{1}{2\cosh(\gamma\hbar B / (2T))}\begin{pmatrix} \exp(-\gamma\hbar B / (2T)) & 0 \\ 0 & \exp(\gamma\hbar B / (2T)) \end{pmatrix}.$$

From which

$$\langle J_x \rangle = \langle J_y \rangle = 0, \langle J_z \rangle = -\frac{\hbar}{2}\tanh(\gamma\hbar B / (2T))$$

The Magnetic Method

Measurements can be made of the Earth's total magnetic field or of components of the field in various directions. The oldest magnetic prospecting instrument is the magnetic compass, which measures the field direction. Other instruments include magnetic balances and fluxgate magnetometres. Most magnetic surveys are made with proton-precession or optical-pumping magnetometres, which are appreciably more accurate. The proton magnetometre measures a radio-frequency voltage induced in a coil by the reorientation (precession) of magnetically polarized protons in a container of ordinary water. The optical-pumping magnetometre makes use of the principles of nuclear resonance and cesium or rubidium vapour. It can detect minute magnetic fluctuations by measuring the effects of light-induced (optically pumped) transitions between atomic energy levels that are dependent on magnetic field strength.

Magnetic surveys are usually made with magnetometres borne by aircraft flying in parallel lines spaced two to four kilometres apart at an elevation of about 500 metres (one metre = 3.28 feet) when exploring for petroleum deposits and in lines 0.5 to one kilometre apart roughly 200 metres above the ground when searching for mineral concentrations. Ground surveys are conducted to follow up magnetic anomaly discoveries made from the air. Such surveys may involve stations spaced only 50 metres apart. Magnetometres also are towed by research vessels. In some cases, two or more magnetometres displaced a few metres from each other are used in a gradiometre arrangement; differences between their readings indicate the magnetic field gradient. A ground monitor is usually used to measure the natural fluctuations of the Earth's field over time so that corrections can be made. Surveying is generally suspended during periods of large magnetic fluctuation (magnetic storms).

Magnetic effects result primarily from the magnetization induced in susceptible rocks by the Earth's magnetic field. Most sedimentary rocks have very low susceptibility and thus are nearly transparent to magnetism. Accordingly, in petroleum exploration magnetics are used negatively: magnetic anomalies indicate the absence of explorable sedimentary rocks. Magnetics are used for mapping features in igneous and metamorphic rocks, possibly faults, dikes, or other features that are associated with mineral concentrations. Data are usually displayed in the form of a contour map of the magnetic field, but interpretation is often made on profiles.

Rocks cannot retain magnetism when the temperature is above the Curie point (about 500° C for most magnetic materials), and this restricts magnetic rocks to the upper 40 kilometres of the Earth's interior. The source of the geomagnetic field must be deeper than this, and it is now believed that convection currents of conducting material in the outer core generate the field. These currents couple to the Earth's spin, so that the magnetic field—when averaged over time—is oriented along the planet's axis. The currents gradually change with time in a somewhat erratic manner and their aggregate effect sometimes reverses, which explains the time changes in the Earth's field. This is the crux of the magnetohydrodynamic theory of the geomagnetic field.

Earth Tide

Earth tide or body tide is the displacement of the solid Earth's surface caused by the gravity of the Moon and Sun. Its main component has metre-level amplitude at periods of about 12 hours and longer. The largest body tide constituents are semi-diurnal, but there are also significant diurnal, semi-annual, and fortnightly contributions. Though the gravitational forcing causing earth tides and ocean tides is the same, the responses are quite different.

Tide Raising Force

The larger of the periodic gravitational forces is from the Moon but that of the Sun is also important. The images here show lunar tidal force when the Moon appears directly over 30° N (or 30° S). This pattern remains fixed with the red area directed toward (or directly away from) the Moon. Red indicates upward pull, blue downward. If, for example the Moon is directly over 90° W (or 90° E), the red areas are centred on the western northern hemisphere, on upper right. Red up, blue down. If for example the Moon is directly over 90° W (90° E), the centre of the red area is 30° N, 90° W and 30° S, 90° E, and the centre of the bluish band follows the great circle equidistant from those points. At 30° latitude a strong peak occurs once per lunar day, giving a significant diurnal force at that latitude. Along the equator two equally sized peaks (and depressions) are equally sized, giving semi-diurnal force there.

Body Tide

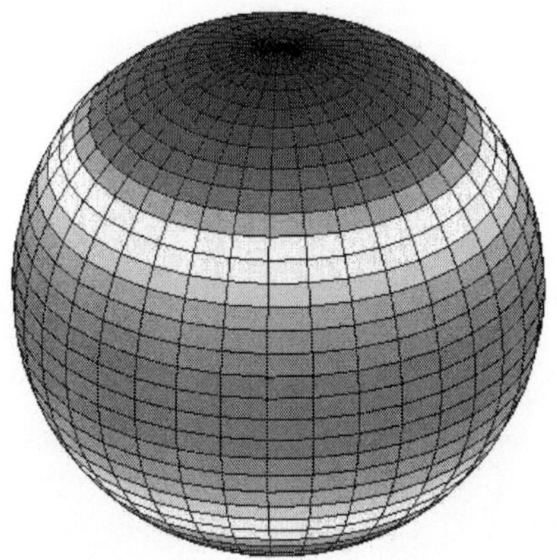

Figure: *Vertical displacements of zonal movement. Red up, blue down.*

The Earth tide encompasses the entire body of the Earth and is unhindered by the thin crust and land masses of the surface, on scales that make the rigidity of rock irrelevant. Ocean tides are a consequence of the resonance of the same driving forces with water movement periods in ocean basins accumulated over many days, so that their amplitude and timing are quite different and vary over short distances of just a few hundred km. The oscillation periods of the earth as a whole are not near the astronomical periods, so its flexing is due to the forces of the moment.

The tide components with a period near twelve hours have a lunar amplitude (earth bulge/depression distances) that are a little more than twice the height of the solar amplitudes, as tabulated below. At new and full moon, the Sun and the Moon are aligned, and the lunar and the solar tidal maxima and minima (bulges and depressions) add together for the greatest tidal range at particular latitudes. At first- and third-quarter phases of the moon, lunar and solar tides are perpendicular, and the tidal range is at a minimum. The semi-diurnal tides go through one full cycle (a high and low tide) about once every 12 hours and one full cycle of maximum height (a spring and neap tide) about once every 14 days.

The classical theory of Earth tides first became established in 1909, primarily to explain nutations, but are also used in Earth rotation predictions. The semi-diurnal tide (one maximum every 12 or so hours) is primarily lunar (only S_2 is purely solar) and gives rise to *sectorial* deformations which rise and fall at the same time along the same longitude. Sectorial variations of vertical and east-west displacements are maximum at the equator and vanish at the poles. There are two cycles along each latitude, the bulges opposite one another, and the depressions similarly opposed.

The diurnal tide is lunisolar, and gives rise to *tesseral* deformations. The vertical and east-west movement is maximum at 45° latitude and is zero on the equator and at the poles. Tesseral variation have one cycle per latitude, one bulge and one depression; the bulges are opposed (antipodal), that is to say the western part of the northern hemisphere and the eastern part of the southern hemisphere, for example, and similarly the depressions are opposed, the eastern part of the northern hemisphere and the western part of the southern hemisphere, in this case. Finally, fortnightly and semi-annual tides have 'zonal' deformations (constant along a circle of latitude), as the Moon or Sun gravitation is directed alternately away from the northern and southern hemispheres due to tilt. There is zero vertical displacement at 35°16' latitude.

Since these displacements affect the vertical direction east-west and north-south variations are often tabulated in milliarcseconds for astronomical use. The vertical displacement is frequent tabulated in μgal, since the gradient of gravity is location dependent so that the distance conversion is only approximately 3 μgal per cm

Other Earth Tide Contributors

In coastal areas because the ocean tide is quite out of step with the earth tide, at high ocean tide there is an excess (or at low tide a deficit) of water about what would be the gravitational equilibrium level and the adjacent ground falls (or rises) in response to the resulting differences in weight. Displacements caused by ocean tidal loading can exceed the displacements due to the earth body tide. Sensitive instruments far inland often have to make similar corrections. Atmospheric loading and storm events may also be measurable, though the masses in movement are less weighty.

Tidal Constituents

Principal tide constituents. The amplitudes may vary from those listed within several per cent.

Semi-diurnal			
Tidal constituent	***Period***	***Vertical amplitude (mm)***	***Horizontal amplitude (mm)***
M_2	12.421 hr	384.83	53.84
S_2 (solar semi-diurnal)	12.000 hr	179.05	25.05
N_2	12.658 hr	73.69	10.31
K_2	11.967 hr	48.72	6.82
Diurnal			
Tidal constituent	***Period***	***Vertical amplitude (mm)***	***Horizontal amplitude (mm)***
K_1	23.934 hr	191.78	32.01
O_1	25.819 hr	158.11	22.05
P_1	24.066 hr	70.88	10.36
φ_1	23.804 hr	3.44	0.43
ψ_1	23.869 hr	2.72	0.21
S_1 (solar diurnal)	24.000 hr	1.65	0.25
Long term			
Tidal constituent	***Period***	***Vertical amplitude (mm)***	***Horizontal amplitude (mm)***
M_f	13.661 days	40.36	5.59
M_m (moon monthly)	27.555 days	21.33	2.96
S_{sa} (solar semi-annual)	0.50000 yr	18.79	2.60
Lunar node	18.613 yr	16.92	2.34
S_a (solar annual)	1.0000 yr	2.97	0.41

Earth Tide Effects

Volcanologists use the regular, predictable Earth tide movements to calibrate and test sensitive volcano deformation monitoring instruments. The tides may also trigger volcanic events. Seismologists have determined that microseismic events are correlated to tidal variations in Central Asia (north of the Himalayas). The semidiurnal amplitude of terrestrial tides can reach about 55 cm at the equator which is important in GPS calibration and VLBI measurements. Also to make precise astronomical angular measurements requires knowledge of the Earth's rate of rotation and nutation, both of which are influenced by earth tides.

Terrestrial tides also need to be taken in account in the case of some particle physics experiments. For instance, at the CERN or SLAC, the very large particle accelerators were designed while taking terrestrial tides into account for proper operation. Among the effects that need to be taken into account are circumference deformation for circular accelerators and particle beam energy. Since tidal forces generate currents of conducting fluids within the interior of the Earth, they affect in turn the Earth's magnetic field itself.

Chapter 6

Applications of the Gravity and Magnetic Methods

Exploring for Oil with Gravity and Magnetics

Discovering and assessing oil or gas deposits requires integration of information culled from geology, geochemistry, drilling, GIS, seismology, EM, potential fields, and other disciplines. While seismic exploration remains the primary method of exploring for petroleum, use of gravity and magnetic methods has continued to expand, based on their contribution to reliable evaluations (and recent discoveries) in deeper, more challenging environments such as sub-salt structures and deep sea.

By gathering geophysical data to narrow the search area within large fields, exploration crews can refine their targets and apply seismic techniques more efficiently. Combining seismic and gravity methods enables oil explorers to better define and focus projects early on, and minimise the risk of conducting expensive investigation before potential is determined. Advances in processing and interpreting gravity and magnetics There was a time when gravity and magnetic surveys were considered the rough first cut to zero in on locations to set up their seismic gear. That's all changing. As the quality of gravity and magnetic data improves, grav/mag methods are being used to complement and constrain traditional seismic data or even used as the only tool.

Applying Gravity and Magnetics in Deepwater Exploration

With its roots in exploration 55 years ago, Petrobras has matured into a world renowned integrated oil company without losing its explorer focus, leading the world in deepwater expertise. Integration of gravity and magnetics as part of their exploratory tools, enabled them to see beneath the ocean's post salt layer, contributing to their Santos Basin discoveries.

Two-Body Problem in General Relativity

The two-body problem (or Kepler problem) in general relativity is the determination of the motion and gravitational field of two bodies as described by the field equations of general relativity. Solving the Kepler problem is essential to calculate the bending of light by gravity and the motion of a planet orbiting its sun. Solutions are also used to describe the motion of binary stars around each other, and estimate their gradual loss of energy through gravitational radiation. It is customary to assume that both bodies are point-like, so that tidal forces and the specifics of their material composition can be neglected.

General relativity describes the gravitational field by curved space-time; the field equations governing this curvature are nonlinear and therefore difficult to solve in a closed form. Only one exact solution, the Schwarzschild solution, has been found for the Kepler problem; this solution pertains when the mass M of one body is overwhelmingly greater than the mass m of the other. If so, the larger mass may be taken as stationary and the sole contributor to the gravitational field. This is a good approximation for a photon passing a star and for a planet orbiting its sun.

The motion of the lighter body (called the "particle" below) can then be determined from the Schwarzschild solution; the motion is a geodesic ("shortest path between two points") in the curved space-time. Such geodesic solutions account for the anomalous precession of the planet Mercury, which is a key piece of evidence supporting the theory of general relativity. They also describe the bending of light in a gravitational field, another prediction famously used as evidence for general relativity.

If both masses are considered to contribute to the gravitational field, as in binary stars, the Kepler problem can be solved only approximately. The earliest approximation method to be developed was the post-Newtonian expansion, an iterative method in which an initial solution is gradually corrected. More recently, it has become possible to solve Einstein's field equation using a computer instead of mathematical formulae. As the two bodies orbit each other, they will emit gravitational radiation; this causes them to lose energy and angular momentum gradually, as illustrated by the binary pulsar PSR B1913+16.

Historical Context

Classical Kepler Problem

The Kepler problem derives its name from Johannes Kepler, who worked as an assistant to the Danish astronomer Tycho Brahe. Brahe took extraordinarily accurate measurements of the motion of the planets of the Solar System. From these measurements, Kepler was able to formulate Kepler's laws, the first modern description of planetary motion:

1. The orbit of every planet is an ellipse with the Sun at one of the two foci.
2. A line joining a planet and the Sun sweeps out equal areas during equal intervals of time.
3. The square of the orbital period of a planet is directly proportional to the cube of the semi-major axis of its orbit.

Kepler published the first two laws in 1609 and the third law in 1619. They supplanted earlier models of the Solar System, such as those of Ptolemy and Copernicus. Kepler's laws apply only in the limited case of the two-body problem. Voltaire and Émilie du Châtelet were the first to call them "Kepler's laws".

Nearly a century later, Isaac Newton had formulated his three laws of motion. In particular, Newton's second law states that a force F applied to a mass m produces an acceleration a given by the equation $F=ma$. Newton then posed the question: what must the force be that produces the elliptical orbits seen by Kepler? His answer came in his law of universal gravitation, which states that the force between a mass M and another mass m is given by the formula

$$F = G\frac{Mm}{r^2},$$

where r is the distance between the masses and G is the gravitational constant. Given this force law and his equations of motion, Newton was able to show that two point masses attracting each other would each follow perfectly elliptical orbits. The ratio of sizes of these ellipses is m/M, with the larger mass moving on a smaller ellipse. If M is much larger than m, then the larger mass will appear to be stationary at the focus of the elliptical orbit of the lighter mass m. This model can be applied approximately to the Solar System. Since the mass of the Sun is much larger than those of the planets, the force acting on each planet is principally due to the Sun; the gravity of the planets for each other can be neglected to first approximation.

Apsidal Precession

In celestial mechanics, perihelion precession, apsidal precession or orbital precession is the precession (rotation) of the orbit of a celestial body. More precisely it is the gradual rotation of the line joining the apsides of an orbit, which are the points of closest and farthest approach. Perihelion is the closest point to the Sun. The apsidal precession is the first derivative of the argument of periapsis, one of the six primary orbital elements of an orbit.

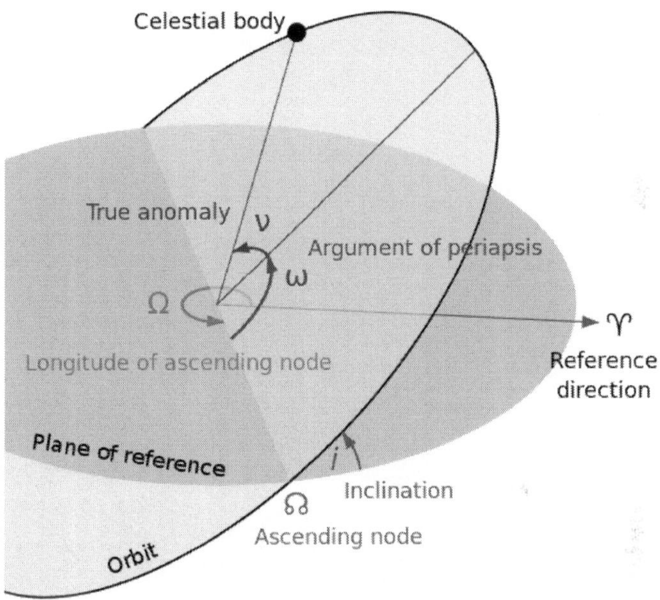

Figure: *The orbital parameters. The line of apsides is shown in blue, and denoted by ω. The apsidal precession is the rate of change of ω through time (dω/dt).*

Calculation

There are a variety of factors which can lead to periastron precession, such as general relativity, stellar quadrupole moments, mutual star–planet tidal deformations, and perturbations from other planets.

$$\omega_{\text{total}} = \omega_{\text{General Relativity}} + \omega_{\text{quadrupole}} + \omega_{\text{tide}} + \omega_{\text{perturbations}}$$

For Mercury, the perihelion precession rate due to general relativistic effects is 433 per century. By comparison, the precession due to perturbations from the other solar system planets is 5323 per century while the oblateness of the Sun (quadrupole moment) causes a negligible contribution of 0.0253 per century.

From classical mechanics, if stars and planets are considered to be purely spherical masses, then they will obey a simple r^{-2} force law and hence execute closed elliptical orbits.

Non-spherical mass effects are caused by the application of external potential(s): the centrifugal potential of spinning bodies causes rotational flattening and the tidal potential of a nearby mass raises tidal bulges. Rotational and tidal bulges create gravitational quadrupole fields (r^{-3}) that lead to orbital precession

Total apsidal precession broadly in order of importance for isolated very hot Jupiters is (considering only lowest order effects)

$$\omega_{\text{total}} = \omega_{\text{tidal perturbations}} + \omega_{\text{General Relativity}} + \omega_{\text{rotational perturbations}} + \omega_{\text{rotational *}} + \omega_{\text{tidal *}}$$

with planetary tidal bulge being the dominant term, exceeding the effects of general relativity and the stellar quadrupole by more than an order of magnitude and thus can help us in understanding their interiors. For the shortest-period planets, the planetary interior induces precession of a few degrees per year and up to 19ζ%.9° per year for WASP-12b.

Newton's Theorem of Revolving Orbits

Newton derived an intriguing theorem showing that variations in the angular motion of a particle can be accounted for by the addition of a force that varies as the inverse cube of distance, without affecting the radial motion of a particle.

Using a forerunner of the Taylor series, Newton generalized his theorem to all force laws provided that the deviations from circular orbits is small, which is valid for most planets in the Solar System.

However, his theorem did not account for the apsidal precession of the Moon without giving up the inverse-square law of Newton's law of universal gravitation.

Perturbation Theory

The expected rate of apsidal precession can be calculated more accurately using the methods of perturbation theory.

General Relativity

An apsidal precession of the planet Mercury was noted by Urbain Le Verrier in the mid-19th century and accounted for by Einstein's theory of general relativity.

To first approximation, this theory adds a central force that varies as the inverse fourth power of the distance.

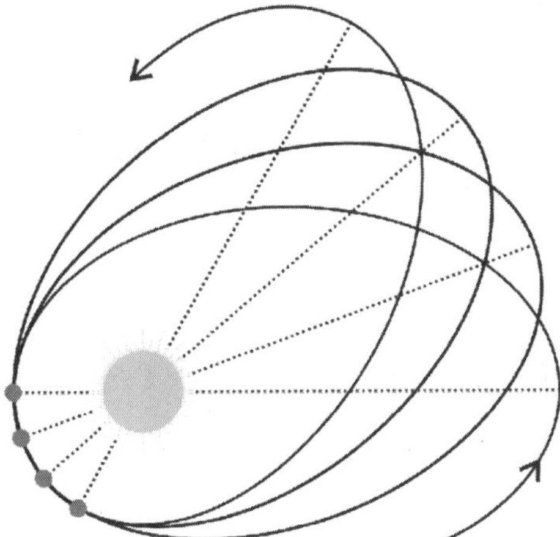

Figure: *Change in orbit over time*

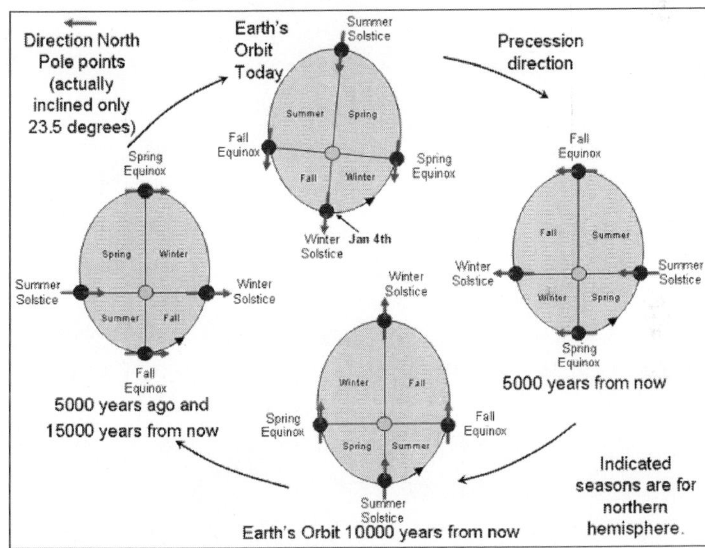

Figure: *Effects of apsidal precession on the seasons*

Einstein showed that for a planet, the major semi-axis of its orbit being α, the eccentricity of the orbit e and the period of revolution T, then the apsidal precession due to relativistic effects, during one period of revolution in radians, is

$$\epsilon = 24\pi^3 \frac{\alpha^2}{T^2 c^2 (1-e^2)}$$

where c is the speed of light. In the case of Mercury, half of the greater axis is circa 57,9 million kilometres or 57,9 $\cdot\ 10^9$ m, the eccentricity of its orbit is 0,206 and the period of revolution 87,97 days, or 7,6 10^6 s.

From these and the speed of light (which is 3 10^8 m/s), it can be calculated that the apsidial precession during one period of revolution is $\epsilon = 5{,}028 \cdot 10^{-7}$ radians, $2{,}88 \cdot 10^{-5}$ degrees or 0,104 arc seconds. In one hundred years, Mercury makes circa 415 revolutions around the Sun, and thus in that time, the apsidal perihelion due to relativistic effects is circa 43 arc seconds, which corresponds almost exactly to the previously unexplained part of the measured value.

Long-Term Climate

Because of apsidal precession the Earth's argument of periapsis slowly increases; it takes about 112,000 years for the ellipse to revolve once relative to the fixed stars. The Earth's polar axis, and hence the solstices and equinoxes, precess with a period of about 26,000 years in relation to the fixed stars. These two forms of 'precession' combine so that it takes about 21,000 years for the ellipse to revolve once relative to the vernal equinox, that is, for the perihelion to return to the same date (given a calendar that tracks the seasons perfectly).

This interaction between the anomalistic and tropical cycle is important in the long-term climate variations on Earth, called the Milankovitch cycles. An equivalent is also known on Mars.

The figure illustrates the effects of precession on the northern hemisphere seasons, relative to perihelion and aphelion. Notice that the areas swept during a specific season changes through time. Orbital mechanics require that the length of the seasons be proportional to the swept areas of the seasonal quadrants, so when the orbital eccentricity is extreme, the seasons on the far side of the orbit may be substantially longer in duration.

Laplace–Runge–Lenz Vector

In classical mechanics, the Laplace–Runge–Lenz vector (or simply the LRL vector) is a vector used chiefly to describe the shape and orientation of the orbit of one astronomical body around another, such as a planet revolving around a star. For two bodies interacting by Newtonian gravity, the LRL vector is a constant of motion, meaning that it is the same no matter where it is calculated on the orbit; equivalently, the LRL vector is said to be *conserved*. More generally, the LRL vector is conserved in all problems in which two bodies interact by a central force that varies as the inverse square of the distance between them; such problems are called Kepler problems.

The hydrogen atom is a Kepler problem, since it comprises two charged particles interacting by Coulomb's law of electrostatics, another inverse square central force. The LRL vector was essential in the first quantum mechanical derivation of the spectrum of the hydrogen atom, before the development of the Schrödinger equation. However, this approach is rarely used today.

In classical and quantum mechanics, conserved quantities generally correspond to a symmetry of the system. The conservation of the LRL vector corresponds to an unusual symmetry; the Kepler problem is mathematically equivalent to a particle moving freely on the surface of a four-dimensional (hyper-)sphere, so that the whole problem is symmetric under certain rotations of the four-dimensional space. This higher symmetry results from two properties of the Kepler problem: the velocity vector always moves in a perfect circle and, for a given total energy, all such velocity circles intersect each other in the same two points.

The Laplace–Runge–Lenz vector is named after Pierre-Simon de Laplace, Carl Runge and Wilhelm Lenz. It is also known as the Laplace vector, the Runge–Lenz vector and the Lenz vector. Ironically, none of those scientists discovered it. The LRL vector has been rediscovered several times and is also equivalent to the dimensionless eccentricity vector of celestial mechanics. Various generalizations of the LRL vector have been defined, which incorporate the effects of special relativity, electromagnetic fields and even different types of central forces.

Context

A single particle moving under any conservative central force has at least four constants of motion, the total energy E and the three Cartesian components of the angular momentum vector L with respect to the origin. The particle's orbit is confined to a plane defined by the particle's initial momentum p (or, equivalently, its velocity v) and the vector r between the particle and the centre of force.

As defined below, the Laplace–Runge–Lenz vector (LRL vector) A always lies in the plane of motion for any central force. However, A is constant only for an inverse-square central force. For most central forces, however, this vector A is not constant, but changes in both length and direction; if the central force is *approximately* an inverse-square law, the vector A is approximately constant in length, but slowly rotates its direction. A *generalized* conserved LRL vector $\mathcal{A}$ can be defined for all central forces, but this generalized vector is a complicated function of position, and usually not expressible in closed form.

The plane of motion is perpendicular to the angular momentum vector L, which is constant; this may be expressed mathematically by the vector dot product equation r ·L = 0; likewise, since A lies in that plane, A ·L = 0.

The LRL vector differs from other conserved quantities in the following property. Whereas for typical conserved quantities, there is a corresponding cyclic coordinate in the three-dimensional Lagrangian of the system, there does *not* exist such a coordinate for the LRL vector. Thus, the conservation of the LRL vector must be derived directly, e.g., by the method of Poisson brackets, as described below. Conserved quantities of this kind are called "dynamic", in contrast to the usual "geometric" conservation laws, e.g., that of the angular momentum.

Anomalous Precession of Mercury

In 1859, Urbain Le Verrier discovered that the orbital precession of the planet Mercury was not quite what it should be; the ellipse of its orbit was rotating (precessing) slightly faster than predicted by the traditional theory of Newtonian gravity, even after all the effects of the other planets had been accounted for. The effect is small (roughly 43 arcseconds of rotation per century), but well above the measurement error (roughly 0.1 arcseconds per century). Le Verrier realised the importance of his discovery immediately, and challenged astronomers and physicists alike to account for it. Several classical explanations were proposed, such as interplanetary dust, unobserved oblateness of the Sun, an undetected moon of Mercury, or a new planet named Vulcan. After these explanations were discounted, some physicists were driven to the more radical hypothesis that Newton's inverse-square law of gravitation was incorrect. For example, some physicists proposed a power law with an exponent that was slightly different from 2.

Others argued that Newton's law should be supplemented with a velocity-dependent potential. However, this implied a conflict with Newtonian celestial dynamics. In his treatise on celestial mechanics, Laplace had shown that if the gravitational influence does not act instantaneously, then the motions of the planets themselves will not exactly conserve momentum (and consequently some of the momentum would have to be ascribed to the mediator of the gravitational interaction, analogous to ascribing momentum to the mediator of the electromagnetic interaction.)

As seen from a Newtonian point of view, if gravitational influence does propagate at a finite speed, then at all points in time a planet is attracted to a point where the Sun was some time before, and not towards the instantaneous position of the Sun. On the assumption of the classical fundamentals, Laplace had shown that if gravity would propagate at a velocity on the order of the speed of light then the solar system would be unstable, and would not exist for a long time. The observation that the solar system is old allows one to put a lower limit on the speed of gravity that is many orders of magnitude faster than the speed of light.

Laplace's estimate for the velocity of gravity is not correct, because in a field theory which respects the principle of relativity, the attraction of a point charge which is moving at a constant velocity is towards the extrapolated instantaneous position, not to the apparent position it seems to occupy when looked at.

To avoid those problems, between 1870 and 1900 many scientists used the electrodynamic laws of Wilhelm Eduard Weber, Carl Friedrich Gauss, Bernhard Riemann to produce stable orbits and to explain the perihelion shift of Mercury's orbit.

In 1890 Lévy succeeded in doing so by combining the laws of Weber and Riemann, whereby the speed of gravity is equal to the speed of light in his theory. And in another attempt Paul Gerber (1898) even succeeded in deriving the correct formula for the perihelion shift (which was identical to that formula later used by Einstein). However, because the basic laws of Weber and others were wrong (for example, Weber's law was superseded by Maxwell's theory), those hypotheses were rejected. Another attempt by Hendrik Lorentz (1900), who already used Maxwell's theory, produced a perihelion shift which was too low.

Einstein's Theory of General Relativity

Figure: *Eddington's 1919 measurements of the bending of star-light by the Sun's gravity led to the acceptance of general relativity worldwide.*

Around 1904–1905, the works of Hendrik Lorentz, Henri Poincaré and finally Albert Einstein's special theory of relativity, exclude the possibility of propagation of any effects faster than the speed of light. It followed that Newton's law of gravitation would have to be replaced with another law, compatible with the principle of relativity, while still obtaining the newtonian limit for circumstances where relativistic effects are negligible. Such attempts were made by Henri Poincaré (1905), Hermann Minkowski (1907) and Arnold Sommerfeld (1910). In 1907 Einstein came to the conclusion that to achieve this a successor to special relativity was needed.

From 1907 to 1915, Einstein worked towards a new theory, using his equivalence principle as a key concept to guide his way. According to this principle, a uniform gravitational field acts equally on everything within it and, therefore, cannot be detected by a free-falling observer. Conversely, all local gravitational effects should be reproducible in a linearly accelerating reference frame, and vice versa.

Thus, gravity acts like a fictitious force such as the centrifugal force or the Coriolis force, which result from being in an accelerated reference frame; all fictitious forces are proportional to the inertial mass, just as gravity is. To effect the reconciliation of gravity and special relativity and to incorporate the equivalence principle, something had to be sacrificed; that something was the long-held classical assumption that our space obeys the laws of Euclidean geometry, e.g., that the Pythagorean theorem is true experimentally.

Einstein used a more general geometry, pseudo-Riemannian geometry, to allow for the curvature of space and time that was necessary for the reconciliation; after eight years of work (1907–1915), he succeeded in discovering the precise way in which space-time should be curved in order to reproduce the physical laws observed in Nature, particularly gravitation. Gravity is distinct from the fictitious forces centrifugal force and coriolis force in the sense that the curvature of spacetime is regarded as physically real, whereas the fictitious forces are not regarded as forces. The very first solutions of his field equations explained the anomalous precession of Mercury and predicted an unusual bending of light, which was confirmed *after* his theory was published. These solutions are explained below.

General Relativity, Special Relativity and Geometry

In the normal Euclidean geometry, triangles obey the Pythagorean theorem, which states that the square distance ds^2 between two points in space is the sum of the squares of its perpendicular components

$$ds^2 = dx^2 + dy^2 + dz^2$$

where dx, dy and dz represent the infinitesimal differences between the x, y and z coordinates of two points in a Cartesian coordinate system. Now imagine a world in which this is not quite true; a world where the distance is instead given by

$$ds^2 = F(x,y,z)dx^2 + G(x,y,z)dy^2 + H(x,y,z)dz^2$$

where F, G and H are arbitrary functions of position. It is not hard to imagine such a world; we live on one. The surface of the world is curved, which is why it's impossible to make a perfectly accurate flat map of the world. Non-Cartesian coordinate systems illustrate this well; for example, in the spherical coordinates (r, θ, φ), the Euclidean distance can be written

$$ds^2 = dr^2 + r^2 d\theta^2 + r^2 \sin^2 \theta d\varphi^2$$

Another illustration would be a world in which the rulers used to measure length were untrustworthy, rulers that changed their length with their position and even their orientation. In the most general case, one must allow for cross-terms when calculating the distance ds

$$ds^2 = g_{xx}dx^2 + g_{xy}dxdy + g_{xz}dxdz + \cdots + g_{zy}dzdy + g_{zz}dz^2$$

where the nine functions g_{xx}, g_{xy}, …, g_{zz} constitute the metric tensor, which defines the geometry of the space in Riemannian geometry. In the spherical-coordinates example above, there are no cross-terms; the only nonzero metric tensor components are $g_{rr} = 1$, $g_{\theta\theta} = r^2$ and $g_{\varphi\varphi} = r^2 \sin^2 \theta$.

In his special theory of relativity, Albert Einstein showed that the distance ds between two spatial points is not constant, but depends on the motion of the observer. However, there is a measure of separation between two points in space-time — called "proper time" and denoted with the symbol $d\tau$ — that *is* invariant; in other words, it doesn't depend on the motion of the observer.

$$c^2 d\tau^2 = c^2 dt^2 - dx^2 - dy^2 - dz^2$$

which may be written in spherical coordinates as

$$c^2 d\tau^2 = c^2 dt^2 - dr^2 - r^2 d\theta^2 - r^2 \sin^2 \theta d\varphi^2$$

This formula is the natural extension of the Pythagorean theorem and similarly holds only when there is no curvature in space-time. In general relativity, however, space and time may have curvature, so this distance formula must be modified to a more general form

$$c^2 d\tau^2 = g_{\mu\nu} dx^\mu dx^\nu$$

just as we generalized the formula to measure distance on the surface of the Earth. The exact form of the metric $g_{\mu\nu}$ depends on the gravitating mass, momentum and energy, as described by the Einstein field equations. Einstein developed those field equations to match the then known laws of Nature; however, they predicted never-before-seen phenomena (such as the bending of light by gravity) that were confirmed later.

Geodesic Equation

According to Einstein's theory of general relativity, particles of negligible mass travel along geodesics in the space-time. In uncurved space-time, far from a source of gravity, these geodesics correspond to straight lines; however, they may deviate from straight lines when the space-time is curved. The equation for the geodesic lines is

$$\frac{d^2 x^\mu}{dq^2} + \Gamma^\mu_{\nu\lambda} \frac{dx^\nu}{dq} \frac{dx^\lambda}{dq} = 0$$

where Γ represents the Christoffel symbol and the variable q parametrizes the particle's path through space-time, its so-called world line. The Christoffel symbol depends only on the metric tensor $g_{\mu\nu}$, or rather on how it changes with position. The variable q is a constant multiple of the proper time τ for timelike orbits (which are travelled by massive particles), and is usually taken to be equal to it. For lightlike (or null) orbits (which are travelled by massless particles such as the photon), the proper time is zero and, strictly speaking, cannot be used as the variable q. Nevertheless, lightlike orbits can be derived as the ultrarelativistic limit of timelike orbits, that is, the limit as the particle mass m goes to zero while holding its total energy fixed.

Schwarzschild Solution

An exact solution to the Einstein field equations is the Schwarzschild metric, which corresponds to the external gravitational field of a stationary, uncharged, non-rotating, spherically symmetric body of mass M. It is characterized by a length scale r_s, known as the Schwarzschild radius, which is defined by the formula

$$r_s = \frac{2GM}{c^2}$$

where G is the gravitational constant. The classical Newtonian theory of gravity is recovered in the limit as the ratio r_s/r goes to zero. In that limit, the metric returns to that defined by special relativity.

In practice, this ratio is almost always extremely small. For example, the Schwarzschild radius r_s of the Earth is roughly 9 mm (^{3}D $_{8}$ inch); at the surface of the Earth, the corrections to Newtonian gravity are only one part in a billion. The Schwarzschild radius of the Sun is much larger, roughly 2953 metres, but at its surface, the ratio r_s/r is roughly 4 parts in a million. A white dwarf star is much denser, but even here the ratio at its surface is roughly 250 parts in a million. The ratio only becomes large close to ultra-dense objects such as neutron stars (where the ratio is roughly 50%) and black holes.

Orbits about the Central Mass

The orbits of a test particle of infinitesimal mass m about the central mass M is given by the equation of motion

$$\left(\frac{dr}{d\tau}\right)^2 = \frac{E^2}{m^2c^2} - \left(1 - \frac{r_s}{r}\right)\left(c^2 + \frac{h^2}{r^2}\right).$$

which can be converted into an equation for the orbit

$$\left(\frac{dr}{d\varphi}\right)^2 = \frac{r^4}{b^2} - \left(1 - \frac{r_s}{r}\right)\left(\frac{r^4}{a^2} + r^2\right)$$

where, for brevity, two length-scales, a and b, have been introduced. They are constants of the motion and depend on the initial conditions (position and velocity) of the test particle. Hence, the solution of the orbit equation is

$$\varphi = \int \frac{1}{r^2}\left[\frac{1}{b^2} - \left(1 - \frac{r_s}{r}\right)\left(\frac{1}{a^2} + \frac{1}{r^2}\right)\right]^{-1/2} dr.$$

Bending of Light by Gravity

The orbit of photons and particles moving close to the speed of light (ultrarelativistic particles) is obtained by taking the limit as the length-scale a goes to infinity. In this limit, the equation for the orbit becomes

$$\varphi = \int \frac{dr}{r^2\sqrt{\dfrac{1}{b^2} - \left(1 - \dfrac{r_s}{r}\right)\dfrac{1}{r^2}}}$$

Expanding in powers of r_s/r, the leading order term in this formula gives the approximate angular deflection $\delta\varphi$ for a massless particle coming in from infinity and going back out to infinity:

$$\delta\varphi \approx \frac{2r_s}{b} = \frac{4GM}{c^2 b}.$$

Here, the length-scale b can be interpreted as the distance of closest approach.

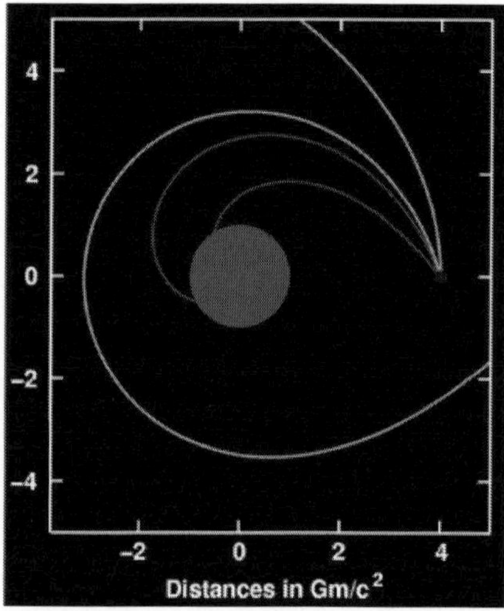

Figure: *Deflection of light (sent out from the location shown in blue) near a compact body (shown in gray)*

Although this formula is approximate, it is accurate for most measurements of gravitational lensing, due to the smallness of the ratio r_s/r. For light grazing the surface of the sun, the approximate angular deflection is roughly 1.75 arcseconds, roughly one millionth part of a circle.

Effective Radial Potential Energy

The equation of motion for the particle derived above

$$\left(\frac{dr}{d\tau}\right)^2 = \frac{E^2}{m^2c^2} - c^2 + \frac{r_s c^2}{r} - \frac{h^2}{r^2} + \frac{r_s h^2}{r^3}$$

can be rewritten using the definition of the Schwarzschild radius r_s as

$$\frac{1}{2}m\left(\frac{dr}{d\tau}\right)^2 = \left[\frac{E^2}{2mc^2} - \frac{1}{2}mc^2\right] + \frac{GMm}{r} - \frac{L^2}{2\mu r^2} + \frac{G(M+m)L^2}{c^2\mu r^3}$$

which is equivalent to a particle moving in a one-dimensional effective potential

$$V(r) = -\frac{GMm}{r} + \frac{L^2}{2\mu r^2} - \frac{G(M+m)L^2}{c^2\mu r^3}$$

The first two terms are well-known classical energies, the first being the attractive Newtonian gravitational potential energy and the second corresponding to the repulsive "centrifugal" potential energy; however, the third term is an attractive energy unique to general relativity. As shown below and elsewhere, this inverse-cubic energy causes elliptical orbits to precess gradually by an angle $\delta\varphi$ per revolution

$$\delta\varphi \approx \frac{6\pi G(M+m)}{c^2 A\left(1-e^2\right)}$$

where A is the semi-major axis and e is the eccentricity.

The third term is attractive and dominates at small r values, giving a critical inner radius r_{inner} at which a particle is drawn inexorably inwards to r=0; this inner radius is a function of the particle's angular momentum per unit mass or, equivalently, the a length-scale defined above.

Circular Orbits and Their Stability

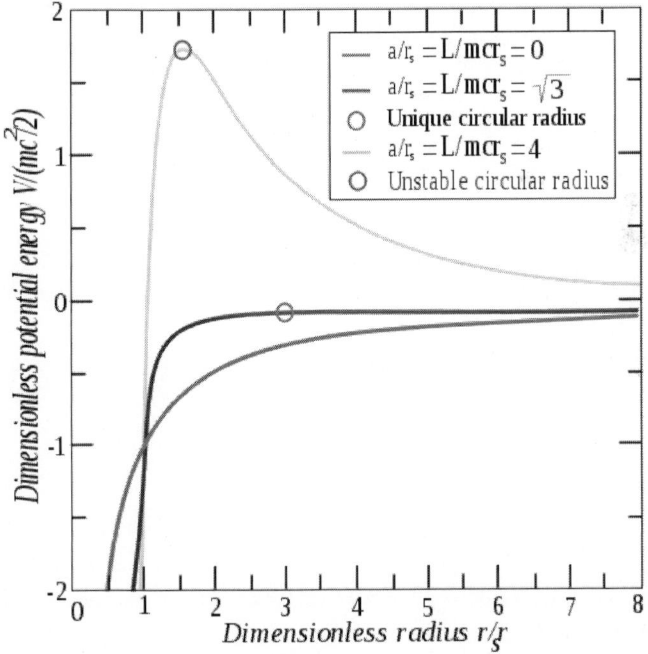

Effective radial potential for various angular momenta. At small radii, the energy drops precipitously, causing the particle to be pulled inexorably inwards to r=0. However, when the normalized angular momentum $a/r_s = L/mcr_s$ equals the square root of three, a metastable circular orbit is possible at the radius highlighted with a green circle.

At higher angular momenta, there is a significant centrifugal barrier (orange curve) and an unstable inner radius, highlighted in red.

The effective potential V can be re-written in terms of the length $a = h/c$:

$$V(r) = \frac{mc^2}{2}\left[-\frac{r_s}{r} + \frac{a^2}{r^2} - \frac{r_s a^2}{r^3}\right].$$

Circular orbits are possible when the effective force is zero:

$$F = -\frac{dV}{dr} = -\frac{mc^2}{2r^4}\left[r_s r^2 - 2a^2 r + 3r_s a^2\right] = 0;$$

i.e., when the two attractive forces—Newtonian gravity (first term) and the attraction unique to general relativity (third term)—are exactly balanced by the repulsive centrifugal force (second term).

There are two radii at which this balancing can occur, denoted here as r_{inner} and r_{outer}:

$$r_{\text{outer}} = \frac{a^2}{r_s}\left(1 + \sqrt{1 - \frac{3r_s^2}{a^2}}\right)$$

$$r_{\text{inner}} = \frac{a^2}{r_s}\left(1 - \sqrt{1 - \frac{3r_s^2}{a^2}}\right) = \frac{3a^2}{r_{\text{outer}}},$$

which are obtained using the quadratic formula. The inner radius r_{inner} is unstable, because the attractive third force strengthens much faster than the other two forces when r becomes small; if the particle slips slightly inwards from r_{inner} (where all three forces are in balance), the third force dominates the other two and draws the particle inexorably inwards to $r = 0$.

At the outer radius, however, the circular orbits are stable; the third term is less important and the system behaves more like the non-relativistic Kepler problem.

When a is much greater than r_s (the classical case), these formulae become approximately

$$r_{\text{outer}} \approx \frac{2a^2}{r_s}$$

$$r_{\text{inner}} \approx \frac{3}{2} r_s$$

Schwarzschild Circular Radii

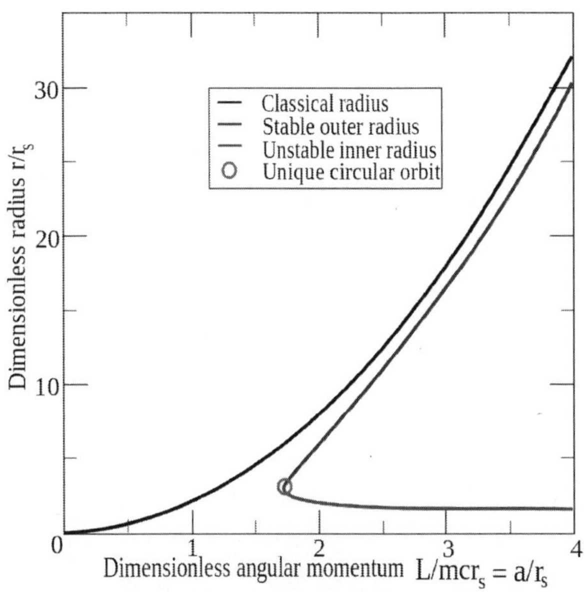

The stable and unstable radii are plotted versus the normalized angular momentum $a/r_s = L/mcr_s$ in blue and red, respectively. These curves meet at a unique circular orbit (green circle) when the normalized angular momentum equals the square root of three. For comparison, the classical radius predicted from the centripetal acceleration and Newton's law of gravity is plotted in black.

Substituting the definitions of a and r_s into r_{outer} yields the classical formula for a particle of mass m orbiting a body of mass M.

$$r_{outer}^3 = \frac{G(M+m)}{\omega_\varphi^2}$$

where ω_φ is the orbital angular speed of the particle. This formula is obtained in non-relativistic mechanics by setting the centrifugal force equal to the Newtonian gravitational force:

$$\frac{GMm}{r^2} = \mu\omega_\varphi^2 r$$

Where μ is the reduced mass.

In our notation, the classical orbital angular speed equals

$$\omega_\varphi^2 \approx \frac{GM}{r_{outer}^3} = \left(\frac{r_s c^2}{2r_{outer}^3}\right) = \left(\frac{r_s c^2}{2}\right)\left(\frac{r_s^3}{8a^6}\right) = \frac{c^2 r_s^4}{16a^6}$$

At the other extreme, when a^2 approaches $3r_s^2$ from above, the two radii converge to a single value

$$r_{\text{outer}} \approx r_{\text{inner}} \approx 3r_s$$

The quadratic solutions above ensure that r_{outer} is always greater than $3r_s$, whereas r_{inner} lies between $^3/_2\, r_s$ and $3r_s$. Circular orbits smaller than $^3/_2\, r_s$ are not possible. For massless particles, a goes to infinity, implying that there is a circular orbit for photons at $r_{\text{inner}} = {}^3/_2\, r_s$. The sphere of this radius is sometimes known as the photon sphere.

Precession of Elliptical Orbits

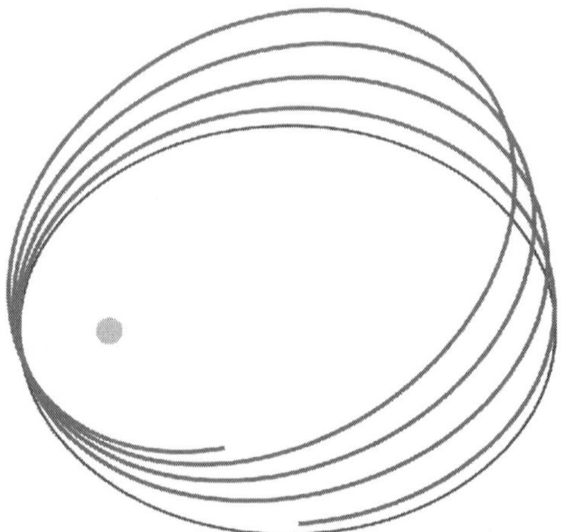

Figure: *In the non-relativistic Kepler problem, a particle follows the same perfect ellipse (red orbit) eternally. General relativity introduces a third force that attracts the particle slightly more strongly than Newtonian gravity, especially at small radii. This third force causes the particle's elliptical orbit to precess (cyan orbit) in the direction of its rotation; this effect has been measured in Mercury, Venus and Earth. The yellow dot within the orbits represents the centre of attraction, such as the Sun.*

The orbital precession rate may be derived using this radial effective potential V. A small radial deviation from a circular orbit of radius r_{outer} will oscillate in a stable manner with an angular frequency

$$\omega_r^2 = \frac{1}{m}\left[\frac{d^2V}{dr^2}\right]_{r=r_{\text{outer}}}$$

which equals

$$\omega_r^2 = \left(\frac{c^2 r_s}{2r_{\text{outer}}^4}\right)\left(r_{\text{outer}} - r_{\text{inner}}\right) = \omega_\varphi^2\sqrt{1-\frac{3r_s^2}{a^2}}$$

Taking the square root of both sides and expanding using the binomial theorem yields the formula

$$\omega_r = \omega_\varphi \left(1 - \frac{3r_s^2}{4a^2} + \cdots \right)$$

Multiplying by the period T of one revolution gives the precession of the orbit per revolution

$$\delta\varphi = T\left(\omega_\varphi - \omega_r\right) \approx 2\pi \left(\frac{3r_s^2}{4a^2} \right) = \frac{3\pi m^2 c^2}{2L^2} r_s^2$$

where we have used $\omega_\varphi T = 2ï$ and the definition of the length-scale a. Substituting the definition of the Schwarzschild radius r_s gives

$$\delta\varphi \approx \frac{3\pi m^2 c^2}{2L^2} \left(\frac{4G^2 M^2}{c^4} \right) = \frac{6\pi G^2 M^2 m^2}{c^2 L^2}$$

This may be simplified using the elliptical orbit's semiaxis A and eccentricity e related by the formula

$$\frac{h^2}{G(M+m)} = A\left(1 - e^2\right)$$

to give the precession angle

$$\delta\varphi \approx \frac{6\pi G(M+m)}{c^2 A\left(1 - e^2\right)}$$

Corrections to the Schwarzschild Solution

Post-Newtonian Expansion: In the Schwarzschild solution, it is assumed that the larger mass M is stationary and it alone determines the gravitational field (i.e., the geometry of space-time) and, hence, the lesser mass m follows a geodesic path through that fixed space-time. This is a reasonable approximation for photons and the orbit of Mercury, which is roughly 6 million times lighter than the Sun. However, it is inadequate for binary stars, in which the masses may be of similar magnitude.

The metric for the case of two comparable masses cannot be solved in closed form and therefore one has to resort to approximation techniques such as the post-Newtonian approximation or numerical approximations. In passing, we mention one particular exception in lower dimensions. In (1+1) dimensions, i.e. a space made of one spatial dimension and one time dimension, the metric for two bodies of equal masses can be solved analytically in terms of the Lambert W function. However, the gravitational energy between the two bodies is exchanged via dilatons rather than gravitons which require three-space in which to propagate.

The post-Newtonian expansion is a calculational method that provides a series of ever more accurate solutions to a given problem. The method is iterative; an initial solution for particle motions is used to calculate the gravitational fields; from these derived fields, new particle motions can be calculated, from which even more accurate estimates of the fields can be computed, and so on. This approach is called "post-Newtonian" because the Newtonian solution for the particle orbits is often used as the initial solution.

When this method is applied to the two-body problem without restriction on their masses, the result is remarkably simple. To the lowest order, the relative motion of the two particles is equivalent to the motion of an infinitesimal particle in the field of their combined masses. In other words, the Schwarzschild solution can be applied, provided that the $M + m$ is used in place of M in the formulae for the Schwarzschild radius r_s and the precession angle per revolution $\delta\varphi$.

Modern Computational Approaches

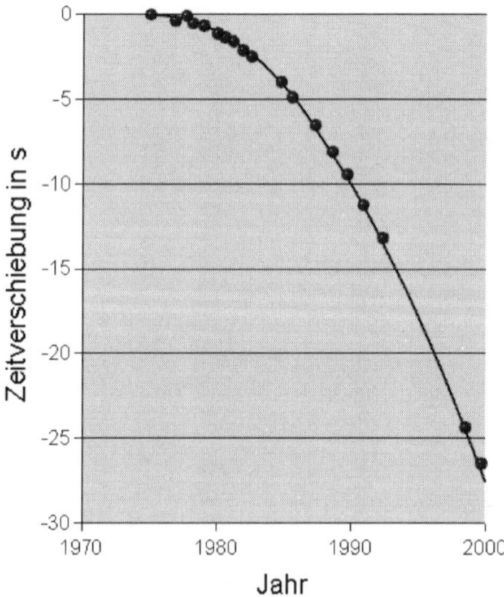

Figure: *Experimentally observed decreases of the orbital period of the binary pulsar PSR B1913+16 (blue dots) match the predictions of general relativity (black curve) almost exactly.*

Einstein's equations can also be solved on a computer using sophisticated numerical methods. Given sufficient computer power, such solutions can be more accurate than post-Newtonian solutions. However, such calculations are demanding because the equations must generally be solved in a four-dimensional space.

Nevertheless, beginning in the late 1990s, it became possible to solve difficult problems such as the merger of two black holes, which is a very difficult version of the Kepler problem in general relativity.

Gravitational Radiation

If there is no incoming gravitational radiation, according to general relativity, two bodies revolving about one another will emit gravitational radiation, causing the orbits to gradually lose energy. This has been observed indirectly in a binary star system known as PSR B1913+16, for which Russell Alan Hulse and Joseph Hooton Taylor, Jr. were awarded the 1993 Nobel Prize in Physics.

The two neutron stars of this system are extremely close and rotate about one another very quickly, completing a revolution in roughly 465 minutes. Their orbit is highly elliptical, with an eccentricity of 0.62 (62%). According to general relativity, the short orbital period and high eccentricity should make the system an excellent emitter of gravitational radiation, thereby losing energy and decreasing the orbital period still further.

The observed decrease in the orbital period over thirty years matches the predictions of general relativity within even the most precise measurements. General relativity predicts that, in another 300 million years, these two stars will spiral into one another.

The formulae describing the loss of energy and angular momentum due to gravitational radiation from the two bodies of the Kepler problem have been calculated. The rate of losing energy (averaged over a complete orbit) is given by

$$-\left\langle \frac{dE}{dt} \right\rangle = \frac{32G^4 m_1^2 m_2^2 \left(m_1 + m_2\right)}{5c^5 a^5 \left(1-e^2\right)^{7/2}} \left(1 + \frac{73}{24}e^2 + \frac{37}{96}e^4\right)$$

where e is the orbital eccentricity and a is the semimajor axis of the elliptical orbit. The angular brackets on the left-hand side of the equation represent the averaging over a single orbit. Similarly, the average rate of losing angular momentum equals

$$-\left\langle \frac{dL_z}{dt} \right\rangle = \frac{32G^{7/2} m_1^2 m_2^2 \sqrt{m_1 + m_2}}{5c^5 a^{7/2} \left(1-e^2\right)^2} \left(1 + \frac{7}{8}e^2\right)$$

The rate of period decrease is given by

$$-\left\langle\frac{dP_b}{dt}\right\rangle = \frac{192G^{5/3}m_1m_2\left(m_1+m_2\right)^{-1/3}}{5c^5\left(1-e^2\right)^{7/2}}\left(1+\frac{73}{24}e^2+\frac{37}{96}e^4\right)\left(\frac{P_b}{2\pi}\right)^{-5/3}$$

where P_b is orbital period.

The losses in energy and angular momentum increase significantly as the eccentricity approaches one, i.e., as the ellipse of the orbit becomes ever more elongated. The radiation losses also increase significantly with a decreasing size *a* of the orbit.

Magnetic Levitation

Magnetic levitation, maglev, or magnetic suspension is a method by which an object is suspended with no support other than magnetic fields. Magnetic force is used to counteract the effects of the gravitational and any other accelerations.

The two primary issues involved in magnetic levitation are *lifting force*: providing an upward force sufficient to counteract gravity, and *stability*: insuring that the system does not spontaneously slide or flip into a configuration where the lift is neutralized.

Magnetic levitation is used for maglev trains, contactless melting, magnetic bearings and for product display purposes.

Lift

Figure: *A superconductor levitating a permanent magnet*

Magnetic materials and systems are able to attract or press each other apart or together with a force dependent on the magnetic field and the area of the magnets, For example, the simplest example of lift would be a simple dipole magnet positioned in the magnetic field of another dipole magnet, oriented with like poles facing each other, so that the force between magnets repels the two magnets.

Essentially all types of magnets have been used to generate lift for magnetic levitation; permanent magnets, electromagnets, ferromagnetism, diamagnetism, superconducting magnets and magnetism due to induced currents in conductors.

To calculate the amount of lift, a magnetic pressure can be defined.

For example, the magnetic pressure of a magnetic field on a superconductor can be calculated by:

$$P_{mag} = \frac{B^2}{2\mu_0}$$

where P_{mag} is the force per unit area in pascals, B is the magnetic field just above the superconductor in teslas, and $\mu_0 = 4\pi\times10^{-7}$ N·A^{-2} is the permeability of the vacuum.

Stability

Earnshaw's theorem proves that using only paramagnetic materials (such as ferromagnetic iron) it is impossible for a static system to stably levitate against gravity.

For example, the simplest example of lift with two simple dipole magnets repelling is highly unstable, since the top magnet can slide sideways, or flip over, and it turns out that no configuration of magnets can produce stability.

However, servomechanisms, the use of diamagnetic materials, superconduction, or systems involving eddy currents allow stability to be achieved.

In some cases the lifting force is provided by magnetic levitation, but stability is provided by a mechanical support bearing little load. This is termed *pseudo-levitation.*

Static

Static stability means that any small displacement away from a stable equilibrium causes a net force to push it back to the equilibrium point.

Earnshaw's theorem proved conclusively that it is not possible to levitate stably using only static, macroscopic, paramagnetic fields. The forces acting on any paramagnetic object in any combinations of gravitational, electrostatic, and magnetostatic fields will make the object's position, at best, unstable along at least one axis, and it can be unstable equilibrium along all axes.

However, several possibilities exist to make levitation viable, for example, the use of electronic stabilization or diamagnetic materials (since relative magnetic permeability is less than one); it can be shown that diamagnetic materials are stable along at least one axis, and can be stable along all axes.

Conductors can have a relative permeability to alternating magnetic fields of below one, so some configurations using simple AC driven electromagnets are self stable.

Dynamic Stability

Dynamic stability occurs when the levitation system is able to damp out any vibration-like motion that may occur.

Magnetic fields are conservative forces and therefore in principle have no built-in damping, and in practice many of the levitation schemes are under-damped and in some cases negatively damped. This can permit vibration modes to exist that can cause the item to leave the stable region.

Damping of Motion is Done in a Number of Ways:

- external mechanical damping (in the support), such as dashpots, air drag etc.
- eddy current damping (conductive metal influenced by field)
- tuned mass dampers in the levitated object
- electromagnets controlled by electronics

Methods

For successful levitation and control of all 6 axes (degrees of freedom; 3 translational and 3 rotational) a combination of permanent magnets and electromagnets or diamagnets or superconductors as well as attractive and repulsive fields can be used. From Earnshaw's theorem at least one stable axis must be present for the system to levitate successfully, but the other axes can be stabilized using ferromagnetism.

The primary ones used in maglev trains are servo-stabilized electromagnetic suspension (EMS), electrodynamic suspension (EDS).

Figure: *Mechanical constraint (in this case the lateral restrictions created by a box) can permit pseudo-levitation of permanent magnets*

Mechanical Constraint (Pseudo-Levitation)

With a small amount of mechanical constraint for stability, achieving pseudo-levitation is a relatively straightforward process.

If two magnets are mechanically constrained along a single vertical axis, for example, and arranged to repel each other strongly, this will act to levitate one of the magnets above the other.

Another geometry is where the magnets are attracted, but constrained from touching by a tensile member, such as a string or cable.

Another example is the Zippe-type centrifuge where a cylinder is suspended under an attractive magnet, and stabilized by a needle bearing from below.

Servomechanisms

The attraction from a fixed strength magnet decreases with increased distance, and increases at closer distances. This is unstable. For a stable system, the opposite is needed, variations from a stable position should push it back to the target position.

Stable magnetic levitation can be achieved by measuring the position and speed of the object being levitated, and using a feedback loop which continuously adjusts one or more electromagnets to correct the object's motion, thus forming a servomechanism.

Figure: *The Transrapid system uses servomechanisms to pull the train up from underneath the track and maintains a constant gap while travelling at high speed*

Many systems use magnetic attraction pulling upwards against gravity for these kinds of systems as this gives some inherent lateral stability, but some use a combination of magnetic attraction and magnetic repulsion to push upwards.

Either system represents examples of ElectroMagnetic Suspension (EMS). For a very simple example, some tabletop levitation demonstrations use this principle, and the object cuts a beam of light to measure the position of the object. The electromagnet is above the object being levitated; the electromagnet is turned off whenever the object gets too close, and turned back on when it falls further away. Such a simple system is not very robust; far more effective control systems exist, but this illustrates the basic idea.

EMS magnetic levitation trains are based on this kind of levitation: The train wraps around the track, and is pulled upwards from below. The servo controls keep it safely at a constant distance from the track.

Induced Currents

These schemes work due to repulsion due to Lenz's law. When a conductor is presented with a time-varying magnetic field electrical currents in the conductor are set up which create a magnetic field that causes a repulsive effect.

These kinds of systems typically show an inherent stability, although extra damping is sometimes required.

Relative Motion between Conductors and Magnets

If one moves a base made of a very good electrical conductor such as copper, aluminium or silver close to a magnet, an (eddy) current will be induced in the conductor that will oppose the changes in the field and create an opposite field that will repel the magnet (Lenz's law). At a sufficiently high rate of movement, a suspended magnet will levitate on the metal, or vice versa with suspended metal. Litz wire made of wire thinner than the skin depth for the frequencies seen by the metal works much more efficiently than solid conductors.

An especially technologically interesting case of this comes when one uses a Halbach array instead of a single pole permanent magnet, as this almost doubles the field strength, which in turn almost doubles the strength of the eddy currents. The net effect is to more than triple the lift force. Using two opposed Halbach arrays increases the field even further.

Halbach arrays are also well-suited to magnetic levitation and stabilisation of gyroscopes and electric motor and generator spindles.

Oscillating Electromagnetic Fields

A conductor can be levitated above an electromagnet (or vice versa) with an alternating current flowing through it. This causes any regular conductor to behave like a diamagnet, due to the eddy currents generated in the conductor. Since the eddy currents create their own fields which oppose the magnetic field, the conductive object is repelled from the electromagnet, and most of the field lines of the magnetic field will no longer penetrate the conductive object.

This effect requires non-ferromagnetic but highly conductive materials like aluminium or copper, as the ferromagnetic ones are also strongly attracted to the electromagnet (although at high frequencies the field can still be expelled) and tend to have a higher resistivity giving lower eddy currents. Again, litz wire gives the best results.

The effect can be used for stunts such as levitating a telephone book by concealing an aluminium plate within it.

At high frequencies (a few tens of kilohertz or so) and kilowatt powers small quantities of metals can be levitated and melted using levitation melting without the risk of the metal being contaminated by the crucible.

One source of oscillating magnetic field that is used is the linear induction motor. This can be used to levitate as well as provide propulsion.

Diamagnetically Stabilized Levitation

Earnshaw's theorem does not apply to diamagnets. These behave in the opposite manner to normal magnets owing to their relative permeability of $\mu_r < 1$ (i.e. negative magnetic susceptibility). Diamagnetic levitation can be inherently stable.

A permanent magnet can be stably suspended by various configurations of strong permanent magnets and strong diamagnets. When using superconducting magnets, the levitation of a permanent magnet can even be stabilized by the small diamagnetism of water in human fingers.

Diamagnetic Levitation

Figure: *Diamagnetic levitation of pyrolytic carbon*

Diamagnetism is the property of an object which causes it to create a magnetic field in opposition to an externally applied magnetic field, thus causing the material to be repelled by magnetic fields. Diamagnetic materials cause lines of magnetic flux to curve away from the material. Specifically, an external magnetic field alters the orbital velocity of electrons around their nuclei, thus changing the magnetic dipole moment. According to Lenz's law, this opposes the external field. Diamagnets are materials with a magnetic permeability less than μ_0 (a relative permeability less than 1). Consequently, diamagnetism is a form of magnetism that is only exhibited by a substance in the presence of an externally applied magnetic field. It is generally quite a weak effect in most materials, although superconductors exhibit a strong effect.

Direct Diamagnetic Levitation

A substance that is diamagnetic repels a magnetic field. All materials have diamagnetic properties, but the effect is very weak, and is usually overcome by the object's paramagnetic or ferromagnetic properties, which act in the opposite manner.

Any material in which the diamagnetic component is strongest will be repelled by a magnet.

Diamagnetic levitation can be used to levitate very light pieces of pyrolytic graphite or bismuth above a moderately strong permanent magnet.

As water is predominantly diamagnetic, this technique has been used to levitate water droplets and even live animals, such as a grasshopper, frog and a mouse.

However, the magnetic fields required for this are very high, typically in the range of 16 teslas, and therefore create significant problems if ferromagnetic materials are nearby.

The minimum criterion for diamagnetic levitation is

$$B\frac{dB}{dz} = \mu_0 \rho \frac{g}{\chi},$$

where:

- χ is the magnetic susceptibility
- ρ is the density of the material
- g is the local gravitational acceleration ("9.8 m/s^2 on Earth)
- μ_0 is the permeability of free space
- $B$is the magnetic field
- $\frac{dB}{dz}$ is the rate of change of the magnetic field along the vertical axis.

Assuming ideal conditions along the z-direction of solenoid magnet:

- Water levitates at $B\frac{dB}{dz} \approx 1400\ \mathrm{T}^2/\mathrm{m}$
- Graphite levitates at $B\frac{dB}{dz} \approx 375\ \mathrm{T}^2/\mathrm{m}$.

Superconductors

Superconductors may be considered perfect diamagnets, and completely expel magnetic fields due to the Meissner effect when the superconductivity initially forms; thus superconducting levitation can be considered a particular instance of diamagnetic levitation. In a type-II superconductor, the levitation of the magnet is further stabilized due to flux pinning within the superconductor; this tends to stop the superconductor from moving with respect to the magnetic field, even if the levitated system is inverted.

These principles are exploited by EDS (Electrodynamic Suspension), superconducting bearings, flywheels, etc.

A very strong magnetic field is required to levitate a train. The JR–Maglev trains have superconducting magnetic coils, but the JR–Maglev levitation is not due to the Meissner effect.

Rotational Stabilization

A magnet with a toroidal field can be stably levitated against gravity when gyroscopically stabilized by spinning it in a second toroidal field created by a base ring of magnet(s). However, this only works while the rate of precession is between both upper and lower critical thresholds—the region of stability is quite narrow both spatially and in the required rate of precession. The first discovery of this phenomenon was by Roy M. Harrigan, a Vermont inventor who patented a levitation device in 1983 based upon it. Several devices using rotational stabilization (such as the popular *Levitron* branded levitating top toy) have been developed citing this patent. Non-commercial devices have been created for university research laboratories, generally using magnets too powerful for safe public interaction.

Strong Focusing

Earnshaw's theory strictly only applies to static fields. Alternating magnetic fields, even purely alternating attractive fields, can induce stability and confine a trajectory through a magnetic field to give a levitation effect. This is used in particle accelerators to confine and lift charged particles, and has been proposed for maglev trains as well.

Uses

Maglev Transportation

Maglev, or magnetic levitation, is a system of transportation that suspends, guides and propels vehicles, predominantly trains, using magnetic levitation from a very large number of magnets for lift and propulsion. This method has the potential to be faster, quieter and smoother than wheeled mass transit systems. The technology has the potential to exceed 6,400 km/h (4,000 mi/h) if deployed in an evacuated tunnel. If not deployed in an evacuated tube the power needed for levitation is usually not a particularly large percentage and most of the power needed is used to overcome air drag, as with any other high speed train.

The highest recorded speed of a maglev train is 581 kilometres per hour (361 mph), achieved in Japan in 2003, 6 km/h faster than the conventional TGV speed record.

Compared to conventional wheeled trains, differences in construction affect the economics of maglev trains. In wheeled trains at very high speeds, the wear and tear from friction along with the hammer effect from wheels on rails accelerates equipment deterioration and prevents mechanically based train systems from routinely achieving higher speeds. Conversely, maglev tracks have historically been found to be much more expensive to construct, but require less maintenance and have lower ongoing costs.

Despite decades of research and development, there are presently only two commercial maglev transport systems in operation, with two others under construction. In April 2004, Shanghai began commercial operations of the high-speed Transrapid system. In March 2005, Japan began operation of the relatively low-speed HSST "Linimo" line in time for the 2005 World Expo.

In its first three months, the Linimo line carried over 10 million passengers. South Korea and the People's Republic of China are both building low-speed maglev lines of their own design, one in Beijing and the other at Seoul's Incheon Airport. Many maglev projects are controversial, and the technological potential, adoption prospects and economics of maglev systems have often been hotly debated. The Shanghai system has been accused of being a white elephant by critics and opponents.

Magnetic Bearings

A magnetic bearing is a bearing that supports a load using magnetic levitation. Magnetic bearings support moving parts without physical contact. For instance, they are able to levitate a rotating shaft and permit relative motion with very low friction and no mechanical wear. Magnetic bearings support the highest speeds of all kinds of bearing and have no maximum relative speed.

Passive magnetic bearings use permanent magnets and, therefore, do not require any input power but are difficult to design due to the limitations described by Earnshaw's theorem. Techniques using diamagnetic materials are relatively undeveloped and strongly depend on material characteristics. As a result, most magnetic bearings are active magnetic bearings, using electromagnets which require continuous power input and an active control system to keep the load stable. In a combined design, permanent magnets are often used to carry the static load and the active magnetic bearing is used when the levitated object deviates from its optimum position. Magnetic bearings typically require a back-up bearing in the case of power or control system failure.

Magnetic bearings are used in several industrial applications such as electrical power generation, petroleum refinement, machine tool operation and natural gas handling. They are also used in the Zippe-type centrifuge, for uranium enrichment and in turbomolecular pumps, where oil-lubricated bearings would be a source of contamination.

Design

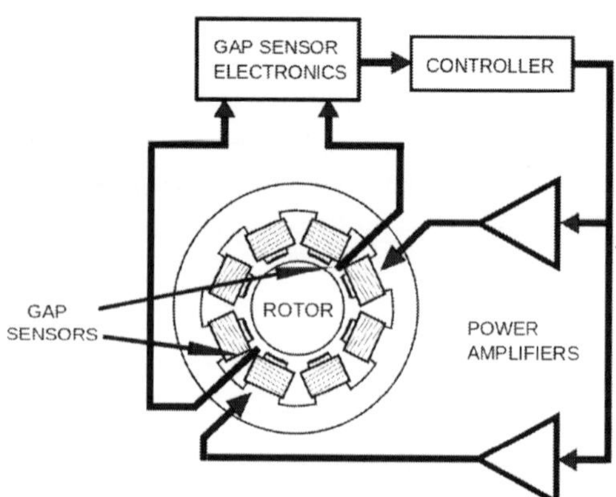

Figure: *Basic operation for a single axis*

An active magnetic bearing works on the principle of electromagnetic suspension and consists of an electromagnet assembly, a set of power amplifiers which supply current to the electromagnets, a controller, and gap sensors with associated electronics to provide the feedback required to control the position of the rotor within the gap. The power amplifier supplies equal bias current to two pairs of electromagnets on opposite sides of a rotor. This constant tug-of-war is mediated by the controller, which offsets the bias current by equal and opposite perturbations of current as the rotor deviates from its centre position.

The gap sensors are usually inductive in nature and sense in a differential mode. The power amplifiers in a modern commercial application are solid state devices which operate in a pulse width modulation configuration. The controller is usually a microprocessor or digital signal processor.

Active bearings have several advantages: they do not suffer from wear, have low friction, and can often accommodate irregularities in the mass distribution automatically, allowing rotors to spin around their centre of mass with very low vibration. Two types of instabilities are typically present in magnetic bearings. Attractive magnets produce an unstable static force that decreases with increasing distance and increases at decreasing distances. This can cause the bearing to become unbalanced. Secondly, because magnetism is a conservative force, it provides little damping; oscillations may cause loss of successful suspension if any driving forces are present.

Applications

Magnetic bearing advantages include very low and predictable friction, and the ability to run without lubrication and in a vacuum. Magnetic bearings are increasingly used in industrial machines such as compressors, turbines, pumps, motors and generators. Magnetic bearings are commonly used in watt-hour metres by electric utilities to measure home power consumption. They are also used in high-precision instruments and to support equipment in a vacuum, for example in flywheel energy storage systems. A flywheel in a vacuum has very low wind resistance losses, but conventional bearings usually fail quickly in a vacuum due to poor lubrication. Magnetic bearings are also used to support maglev trains in order to get low noise and smooth ride by eliminating physical contact surfaces. Disadvantages include high cost, heavy weight and relatively large size.

A new application of magnetic bearings is in artificial hearts. The use of magnetic suspension in ventricular assist devices was pioneered by Prof. Paul Allaire and Prof. Houston Wood at the University of Virginia, culminating in the first magnetically suspended ventricular assist centrifugal pump (VAD) in 1999.

Future Advances

With the use of an induction-based levitation system present in maglev technologies such as the Inductrack system, magnetic bearings could replace complex control systems by using Halbach Arrays and simple closed loop coils. These systems gain in simplicity, but are less advantageous with regard to eddy current losses. For rotating systems it is possible to use homopolar magnet designs instead of multipole Halbach structures, which reduce losses considerably. An example that has bypassed the Earnshaw's theorem issues is the homopolar electrodynamic bearing invented by Dr Torbjφrn Lembke.

This is a novel type of electromagnetic bearing based on a passive magnetic technology. It does not require any control electronics to operate and works because the electrical currents generated by motion cause a restoring force.

Levitation Melting

Electromagnetic levitation (EML), patented by Muck in 1923, is one of the oldest levitation techniques used for containerless experiments. The technique enables the levitation of an object using electromagnets. A typical EML coil has reversed winding of upper and lower sections energized by a radio frequency power supply.

Electrodynamic Suspension

Electrodynamic suspension (EDS) is a form of magnetic levitation in which there are conductors which are exposed to time-varying magnetic fields. This induces eddy currents in the conductors that creates a repulsive magnetic field which holds the two objects apart. These time varying magnetic fields can be caused by relative motion between two objects. I

n many cases, one magnetic field is a permanent field, such as a permanent magnet or a superconducting magnet, and the other magnetic field is induced from the changes of the field that occur as the magnet moves relative to a conductor in the other object.

Electrodynamic suspension can also occur when an electromagnet driven by an AC electrical source produces the changing magnetic field, in some cases, a linear induction motor generates the field.

EDS is used for maglev trains, such as the Japanese SCMaglev. It is also used for some classes of magnetically levitated bearings.

Types

Many examples of this have been used over the years.

Bedford Levitator

In this early configuration by Bedford, Peer, and Tonks from 1939, an aluminum plate is placed on two concentric cylindrical coils, and driven with an AC current. When the parameters are correct, the plate exhibits 6-axis stable levitation.

Levitation Melting

In the 1950s, a technique was developed where small quantities of metal were levitated and melted by a magnetic field of a few tens of kHz. The coil was a metal pipe, allowing coolant to be circulated through it. The overall form was generally conical, with a flat top. This permitted an inert atmosphere to be employed, and was commercially successful.

Linear Induction Motor

Eric Laithwaite and colleagues took the Bedford levitator, and by stages developed and improved it.

First they made the levitator longer along one axis, and were able to make a levitator that was neutrally stable along one axis, and stable along all other axes.

Further development included replacing the single phase energising current with a linear induction motor which combined levitation and thrust.

Later "traverse-flux" systems at his Imperial College laboratory, such as Magnetic river avoided most of the problems of needing to have long, thick iron backing plates when having very long poles, by closing the flux path laterally by arranging the two opposite long poles side-by-side. They were also able to break the levitator primary into convenient sections which made it easier to build and transport.

Null Flux

Null flux systems work by having coils that are exposed to a magnetic field, but are wound in figure of 8 and similar configurations such that when there is relative movement between the magnet and coils, but centred, no current flows since the potential cancels out. When they are displaced off-centre, current flows and a strong field is generated by the coil which tends to restore the spacing.

These schemes were proposed by Powell and Danby in the 1960s, and they suggested that superconducting magnets could be used to generate the high magnetic pressure needed.

Inductrack

Inductrack is a passive, fail-safe magnetic levitation system, using only unpowered loops of wire in the track and permanent magnets (arranged into Halbach arrays) on the vehicle to achieve magnetic levitation. The track can be in one of two configurations, a "ladder track" and a "laminated track". The ladder track is made of unpowered Litz wire cables, and the laminated track is made out of stacked copper or aluminium sheets. There are two designs: the Inductrack I, which is optimized for high speed operation, and the Inductrack II, which is more efficient at lower speeds.

Electrodynamic Bearing

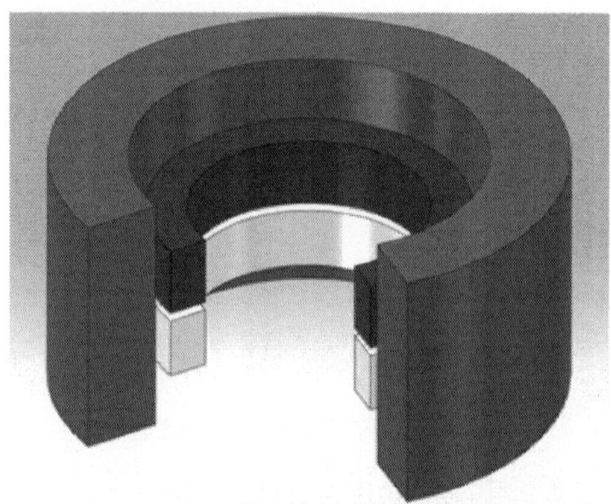

Figure: *3D-image of an axially magnetized ring magnet surrounded by a copper cylinder. The metal ring around the outside spins and the currents generated when it is off-centre relative to the magnet push it back into alignment.*

Electrodynamic bearings (EDB) are a novel type of bearing that is a passive magnetic technology. EDBs do not require any control electronics to operate. They work by the electrical currents generated by motion causing a restoring force.

Uses

Trains

In EDS maglev trains, both the rail and the train exert a magnetic field, and the train is levitated by the repulsive force between these magnetic fields. The magnetic field in the train is produced by either superconducting magnets (as in SCMaglev) or by an array of permanent magnets (as in Inductrack). The repulsive force in the track is created by an induced magnetic field in wires or other conducting strips in the track. A major advantage of the repulsive maglev systems is that they are naturally stable - minor *narrowing* in distance between the track and the magnets creates strong forces to repel the magnets back to their original position, while a slight increase in distance greatly reduces the force and again returns the vehicle to the right separation. No feedback control is necessarily needed.

Repulsive systems have a major downside as well. At slow speeds, the current induced in these coils by the slow change in magnetic flux with respect to time, is not large enough to produce a repulsive electromagnetic force sufficient to support the weight of the train. For this reason the train must have wheels or some other form of landing gear to support the train until it reaches a speed that can sustain levitation. Since a train may stop at any location, due to equipment problems for instance, the entire track must be able to support both low-speed and high-speed operation.

Another downside is that the repulsive system naturally creates a field in the track in front and to the rear of the lift magnets, which act against the magnets and create a form of drag. This is generally only a concern at low speeds, at higher speeds the effect does not have time to build to its full potential and other forms of drag dominate.

The drag force can be used to the electrodynamic system's advantage however, as it creates a varying force in the rails that can be used as a reactionary system to drive the train, without the need for a separate reaction plate, as in most linear motor systems.

Alternatively, propulsion coils on the guideway are used to exert a force on the magnets in the train and make the train move forward. The propulsion coils that exert a force on the train are effectively a linear motor: an alternating current flowing through the coils generates a continuously varying magnetic field that moves forward along the track. The frequency of the alternating current is synchronized to match the speed of the train. The offset between the field exerted by magnets on the train and the applied field creates a force moving the train forward.

Principles

When a conductive loop experiences a changing magnetic field, from Lenz's law and Faraday's law, the changing magnetic field generates an Electromotive Force (EMF) around the circuit. For a sinusoidal excitation, this EMF is 90 degrees phased ahead of the field, peaking where the changes are most rapid (rather than when it is strongest):

$$\mathcal{E} = -N\frac{d\Phi_B}{dt}$$

where N is the number of turns of wire (for a simple loop this is 1) and ϑ_B is the magnetic flux in webers through a *single* loop.

Since the field and potentials are out of phase, both attractive and repulsive forces are produced, and it might be expected that no net lift would be generated. However, although the EMF is at 90 degrees to the applied magnetic field, the loop inevitably has inductance. This inductive impedance tends to delay the peak current, by a phase angle dependent on the frequency (since the inductive impedance of any loop increases with frequency).

$$K = R + i\omega L$$

where K is impedance of the coil, L is the inductance and R is the resistance, the actual phase lead being derivable as the inverse tangent of the product ωL/R, *viz.*, the standard phase lead evidence in a single-loop RL circuit.

But:

$$\mathcal{E} = IK$$

where I is the current.

Thus at low frequencies, the phases are largely orthogonal and the currents lower, and no significant lift is generated. But at sufficiently high frequency, the inductive impedance dominates and the current and the applied field are virtually in line, and this current generates a magnetic field that is opposed to the applied one, and this permits levitation.

However, since the inductive impedance increases proportionally with frequency, so does the EMF, so the current tends to a limit when the resistance is small relative to the inductive impedance. This also limits the lift force. Power used for levitation is therefore largely constant with frequency. However there are also eddy currents due to the finite size of conductors used in the coils, and these continue to grow with frequency.

Since the energy stored in the air gap can be calculated from HB/2 (or $\mu_0 H^2/2$) times air-gap volume, the force applied across the air gap in the direction perpendicular to the load (*viz.*, the force that directly counteracts gravity) is given by the spatial derivative (= gradient) of that energy. The air-gap volume equals the cross-sectional area multiplied by the width of the air gap, so the width cancels out and we are left with a suspensive force of $\mu_0 H^2/2$ times air-gap cross-sectional area, which means that maximum bearable load varies as the square of the magnetic field density of the magnet, permanent or otherwise and varies directly as the cross-sectional area.

Stability

Static: Unlike configurations of simple permanent magnets, electrodynamic levitation can be made stable. Electrodynamic levitation with metallic conductors exhibits a form of diamagnetism, and relative permeabilities of around 0.7 can be achieved (depending on the frequency and conductor configuration). Given the details of the applicable hysteresis loop, frequency-dependent variability of behaviour should be of minimal importance for those magnetic materials that are likely to be deployed.

Dynamic

This form of maglev can cause the levitated object to be subject to a drag induced oscillation, and this oscillation always occurs at a sufficiently high speed. These oscillations can be quite serious and can cause the suspension to fail.

However, inherent system level damping can frequently avoid this from occurring, particularly on large scale systems.

Alternatively, addition of lightweight tuned mass dampers can prevent oscillations from being problematic. Electronic stabilization can also be employed.

Chapter 7

Geomagnetic Pole

The geomagnetic poles are antipodal points where the axis of a best-fitting dipole intersects the Earth's surface. This dipole is equivalent to a powerful bar magnet at the centre of the Earth, and it is this theoretical dipole that comes closer than any other to accounting for the magnetic field observed at the Earth's surface.

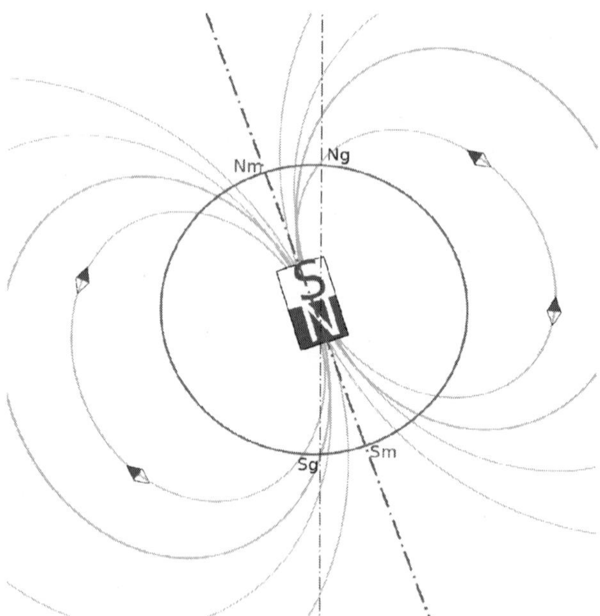

Figure: *Illustration of the difference between geomagnetic poles (Nm and Sm) and geographical poles (Ng and Sg)*

In contrast, the actual magnetic poles are not antipodal—that is, they do not lie on a line passing through the centre of the Earth.

Owing to motion of fluid in the Earth's outer core, the geomagnetic poles are constantly moving. However, over thousands of years their direction averages to the Earth's rotation axis. On the order of once every half a million years, the poles *reverse* (north changes place with south).

Definition

As a first-order approximation, the Earth's magnetic field can be modelled as a simple dipole (like a bar magnet), tilted about 10° with respect to the Earth's rotation axis (which defines the Geographic North and Geographic South Poles) and centred at the Earth's centre.

The North and South Geomagnetic Poles are the antipodal points where the axis of this theoretical dipole intersects the Earth's surface. If the Earth's magnetic field were a perfect dipole then the field lines would be vertical at the Geomagnetic Poles, and they would coincide with the North and South magnetic poles. However, the approximation is imperfect, and so the Magnetic and Geomagnetic Poles lie some distance apart.

Location

Like the North Magnetic Pole, the North Geomagnetic Pole attracts the north pole of a bar magnet and so is in a physical sense actually a *south* magnetic pole. It is the centre of the 'open' magnetic field lines which connect to the interplanetary magnetic field and provide a direct route for the solar wind to reach the ionosphere. As of 2010 it was located at approximately 80.08°N 72.21°W, on Ellesmere Island.

The locations of geomagnetic poles are predicted by the International Geomagnetic Reference Field, a statistical fit to measurements of the Earth's field by satellites and in geomagnetic observatories. If the Earth's field were exactly dipolar, the north pole of a magnetic compass needle would point directly at the North Geomagnetic Pole.

In practice it does not because the geomagnetic field that originates in the core has a more complex non-dipolar part, and magnetic anomalies in the Earth's crust also contribute to the local field.

Movement

The geomagnetic poles move over time because the geomagnetic field is produced by motion of the molten iron alloys in the Earth's outer core. Over the past 150 years the poles have moved westward at a rate of 0.05° to 0.1° per year, with little net north or south motion.

Over several thousand years, the average location of the geomagnetic poles coincides with the geographical poles. Paleomagnetists have long relied on the *Geocentric axial dipole (GAD) hypothesis*, which states that, aside from during geomagnetic reversals, the time-averaged position of the geomagnetic poles has always coincided with the geographic poles. There is considerable paleomagnetic evidence supporting this hypothesis.

North Magnetic Pole

The North Magnetic Pole is the point on the surface of Earth's Northern Hemisphere at which the planet's magnetic field points vertically downwards (in other words, if a magnetic compass needle is allowed to rotate about a horizontal axis, it will point straight down). There is only one location where this occurs, near (but distinct from) the Geographic North Pole and the Geomagnetic North Pole.

The North Magnetic Pole moves over time due to magnetic changes in the Earth's core. In 2001, it was determined by the Geological Survey of Canada to lie near Ellesmere Island in northern Canada at 81.3°N 110.8°W. It was situated at 83.1°N 117.8°W in 2005. In 2009, while still situated within the Canadian Arctic territorial claim at 84.9°N 131.0°W, it was moving toward Russia at between 34 and 37 miles (55 and 60 km) per year. As of 2012, the pole is projected to have moved beyond the Canadian Arctic territorial claim to 85.9°N 147.0°W.

Its southern hemisphere counterpart is the South Magnetic Pole. Since the Earth's magnetic field is not exactly symmetrical, the North and South Magnetic Poles are not antipodal: i.e., a line drawn from one to the other does not pass through the geometric centre of the Earth.

The Earth's North and South Magnetic Poles are also known as Magnetic Dip Poles, with reference to the vertical "dip" of the magnetic field lines at those points.

Polarity

All magnets have two poles, where the lines of magnetic flux enter and emerge. By analogy with the Earth's magnetic field, these are called the magnet's "north" and "south" poles. The convention in early compasses was to call the end of the needle pointing to the Earth's North Magnetic Pole the "north pole" (or "north-seeking pole") and the other end the "south pole" (the names are often abbreviated to "N" and "S"). Because opposite poles attract, this definition means that the Earth's North Magnetic Pole is actually a magnetic *south* pole and the Earth's South Magnetic Pole is a magnetic *north* pole. The direction of magnetic field lines are defined to emerge from the magnet's north pole and enter the magnet's south pole.

History

Early European navigators believed that compass needles were attracted to a "magnetic mountain" or "magnetic island" somewhere in the far north, or to the Pole Star. The idea that the Earth itself acts as a giant magnet was first proposed in 1600 by the English physician and natural philosopher William Gilbert. He was also the first to define the North Magnetic Pole as the point where the Earth's magnetic field points vertically downwards. This is the definition used nowadays, though it would be a few hundred years before the nature of the Earth's magnetic field was understood properly.

Expeditions and Measurements

Early

The first expedition to reach the North Magnetic Pole was led by James Clark Ross, who found it at Cape Adelaide on the Boothia Peninsula on June 1, 1831. Roald Amundsen found the North Magnetic Pole in a slightly different location in 1903. The third observation was by Canadian government scientists Paul Serson and Jack Clark, of the Dominion Astrophysical Observatory, who found the pole at Allen Lake on Prince of Wales Island in 1947.

Project Polaris

At the start of the Cold War, the United States Department of War recognised a need for a comprehensive survey of the North American Arctic and asked the United States Army to undertake the task. An assignment was made in 1946 for the newly formed Army's Air Corps Strategic Air Command to explore the entire Arctic Ocean area. The exploration was conducted by the 46th (later re-designated the 72nd) Photo Reconnaissance Squadron and reported on as a classified *Top Secret* mission named Project Nanook. This project in turn was divided into many separate, but identically classified, projects, one of which was Project Polaris, which was a radar, photographic (trimetrogon, or three-angle, cameras) and visual study of the entire Canadian Archipelago. A Canadian officer observer was assigned to accompany each flight.

Directing Project Polaris was its navigation leader, 1st Lieutenant Frank O. Klein, a World War II combat veteran. Incidental to the project and taken up at his own initiative was a study of northern terrestrial magnetism. The study was prompted by the surprise that the fluxgate compass did not behave erratically as expected. It oscillated no more than 1 to 2 degrees over much of the region. With the cooperation of many of his squadron teammates in obtaining many hundreds of statistical readings, startling results were revealed:

The centre of the north magnetic dip pole was on Prince of Wales Island some 250 miles NNW of the positions determined by Amundsen and Ross, and the dip pole occupied a larger elliptical area, with foci about 250 miles apart on Boothia Peninsula and Bathurst Island.

Klein called the two foci local poles, for their importance to navigation in emergencies when using a "homing" procedure. About 3 months after Klein's findings were officially reported, a Canadian ground expedition was sent into the Archipelago to locate the position of the magnetic pole. R. Glenn Madill, Chief of Terrestrial Magnetism, Department of Mines and Resources, Canada, wrote to Lt. Klein on 21 July 1948:

> *... we agree on one point and that is the presence of what we can call the main magnetic pole on northwestern Prince of Wales Island. I have accepted as a purely preliminary value the position latitude 73°N and longitude 100°W. Your value of 73°15'N and 99°45'W is in excellent agreement, and I suggest that you use your value by all means.*
>
> —R. Glenn Madill

(The positions were less than 20 miles apart.)

Modern (Post 1996)

The Canadian government has made several measurements since, which show that the North Magnetic Pole is moving continually northwestward. In 2001, an expedition located the pole at 81.3°N 110.8°W. In 2007, the latest survey found the pole at 83.95°N 120.72°W. During the 20th century it moved 1100 km, and since 1970 its rate of motion has accelerated from 9 km/year to approximately 52 km/year (2001–2007 average). Members of the 2007 expedition to locate the magnetic north pole wrote that such expeditions have become logistically difficult, as the pole moves further away from inhabited locations. They expect that in the future, the magnetic pole position will be obtained from satellite data instead of ground surveys.

This general movement is in addition to a daily or *diurnal* variation in which the North Magnetic Pole describes a rough ellipse, with a maximum deviation of 80 km from its mean position.

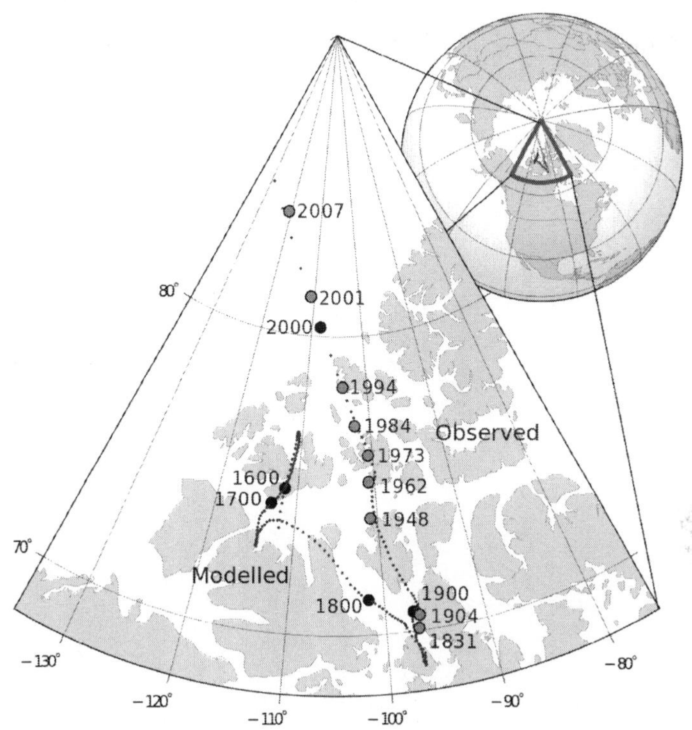

Figure: *The movement of Earth's north magnetic pole across the Canadian arctic, 1831–2007.*

This effect is due to disturbances of the geomagnetic field by charged particles from the Sun.

North Magnetic Pole	(2001) 81.3°N 110.8°W	(2004 est) 82.3°N 113.4°W	(2007) 83.95°N 120.72°W
South Magnetic Pole	(1998) 64.6°S 138.5°E	(2004 est) 63.5°S 138.0°E	(2007) 64.497°S 137.684°E

The first team of novices to reach the Magnetic North Pole did so in 1996, led by David Hempleman-Adams. It included the first British woman Sue Stockdale and first Swedish woman to reach the Pole. The team also successfully tracked the location of the Magnetic North Pole on behalf of the University of Ottawa, and certified its location by magnetometre and theodolite at 78°35.72 N 104°11.92 W.

The biennial Polar Race takes place between Resolute Bay in northern Canada and the 1996-certified location of the North Magnetic Pole at 78°35.72 N 104°11.92 W. On 25 July 2007, the *Top Gear Polar Challenge Special* was broadcast on BBC Two in the United Kingdom, in which Jeremy Clarkson and James May (and their support and camera team) became the first people in history to reach this location in a car.

Arctic Exploration

Arctic exploration is the physical exploration of the Arctic region of the Earth. It refers to the historical period during which mankind has explored the region north of the Arctic Circle. Historical records suggest that humankind have explored the northern extremes since 325 BC, when the ancient Greek sailor Pytheas reached a frozen sea while attempting to find a source of the metal tin. Dangerous oceans and poor weather conditions often fetter explorers attempting to reach polar regions and journeying through these perils by sight, boat, and foot has proven difficult.

Farthest North

Farthest North (sometimes known as Furthest North) describes the most northerly latitude reached by explorers before the conquest of the North Pole rendered the expression obsolete. The northern (Arctic) polar regions are much more accessible than those of the south, as continental land masses extend to high latitudes and sea voyages to the regions are relatively short.

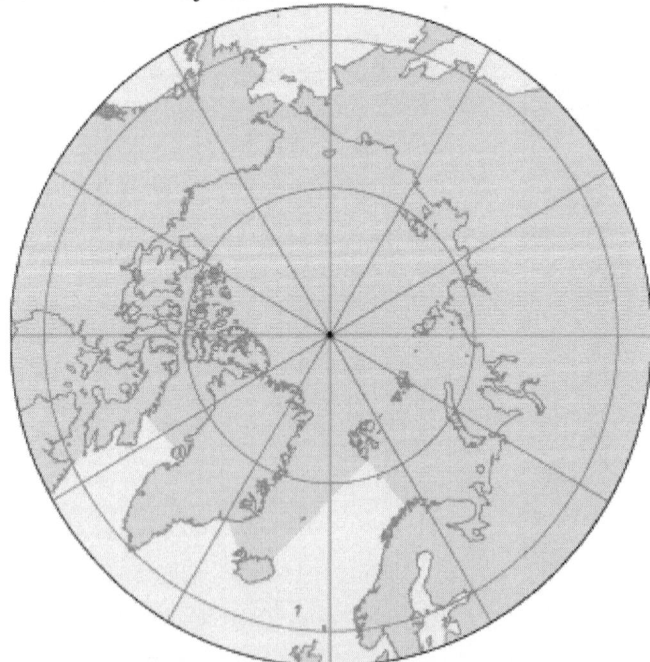

Figure: *The location of the North Pole, at the centre of the Arctic Ocean*

Magnetic North and Magnetic Declination

2000

Declination (degrees east)

http://geomag.usgs.gov

International Geomagnetic Reference Field (IGRF)

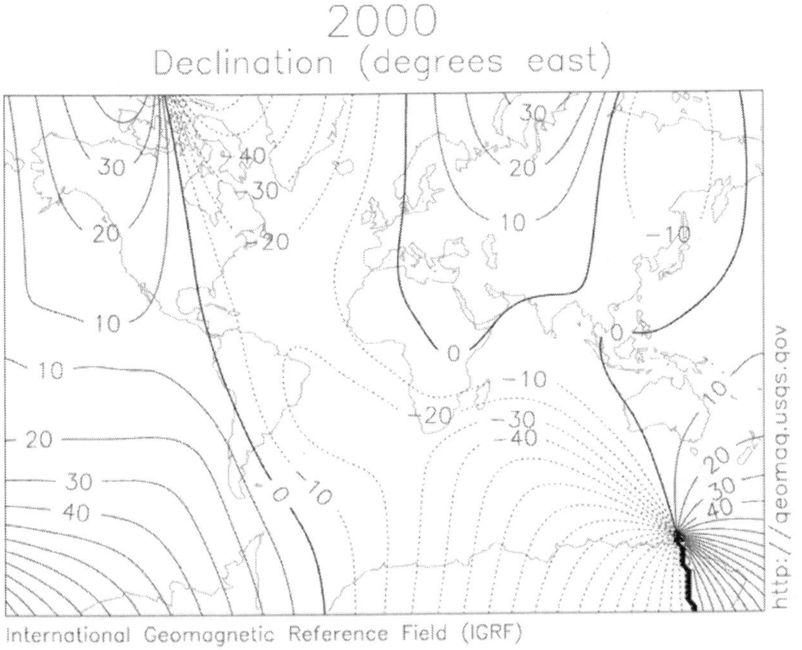

Figure: *Magnetic declination from true north in 2000.*

The direction in which a compass needle points is known as magnetic north. In general, this is not exactly the direction of the North Magnetic Pole (or of any other consistent location). Instead, the compass aligns itself to the local geomagnetic field, which varies in a complex manner over the Earth's surface, as well as over time. The local angular difference between magnetic north and true north is called the magnetic declination. Most map coordinate systems are based on true north, and magnetic declination is often shown on map legends so that the direction of true north can be determined from north as indicated by a compass.

Magnetic declination has been measured in many countries, including the U.S. The line of zero declination (the *agonic line*) in North America runs from the North Magnetic Pole through Lake Superior and southward into the Gulf of Mexico. Along this line, true north is the same as magnetic north. West of the line of zero declination, a compass will give a reading that is east of true north. Conversely, east of the line of zero declination, a compass reading will be west of true north.

Magnetic declination is still very important for certain types of navigation that have traditionally made much use of magnetic compasses.

North Geomagnetic Pole

As a first-order approximation, the Earth's magnetic field can be modelled as a simple dipole (like a bar magnet), tilted about 10° with respect to the Earth's rotation axis (which defines the Geographic North and Geographic South Poles) and centred at the Earth's centre. The North and South Geomagnetic Poles are the antipodal points where the axis of this theoretical dipole intersects the Earth's surface. If the Earth's magnetic field were a perfect dipole then the field lines would be vertical at the Geomagnetic Poles, and they would coincide with the Magnetic Poles. However, the approximation is imperfect, and so the Magnetic and Geomagnetic Poles lie some distance apart.

Like the North Magnetic Pole, the North Geomagnetic Pole attracts the north pole of a bar magnet and so is in a physical sense actually a *south* magnetic pole. It is the centre of the region of the magnetosphere in which the Aurora Borealis can be seen. As of 2005 it was located at approximately 79.74°N 71.78°W, off the northwest coast of Greenland, but it is now drifting away from North America and toward Siberia.

South Magnetic Pole

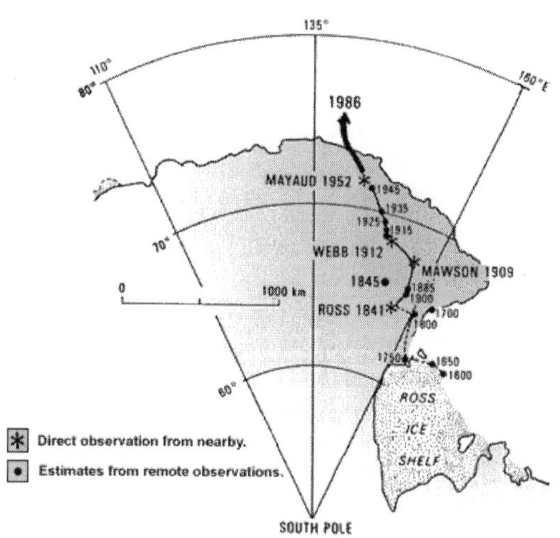

Figure: *Locations of South Magnetic Pole from direct observation and model prediction.*

The South Magnetic Pole is the wandering point on the Earth's Southern Hemisphere where the geomagnetic field lines are directed vertically upwards. It should not be confused with the lesser known South Geomagnetic Pole described later.

For historical reasons, the "end" of a magnet that points (roughly) north is itself called the "north pole" of the magnet, and the other end, pointing south, is called magnet's "south pole". Because opposite poles attract, the Earth's South Magnetic Pole is physically actually a magnetic *north* pole.

The South Magnetic Pole is constantly shifting due to changes in the Earth's magnetic field. As of 2005 it was calculated to lie at 64°312 483 S 137°512 363 E, placing it off the coast of Antarctica, between Adelie Land and Wilkes Land. That point lies outside the Antarctic Circle. Due to polar drift, the pole is moving northwest by about 10 to 15 kilometres per year. Its current distance from the actual Geographic South Pole is approximately 2860 km. The nearest permanent science station is Dumont d'Urville Station. Wilkes Land contains a large gravitational mass concentration.

South Geomagnetic Pole

The Earth's geomagnetic field can be approximated by a tilted dipole (like a bar magnet) placed at the centre of the Earth. The South Geomagnetic Pole is the point where the axis of this best-fitting tilted dipole intersects the Earth's surface in the southern hemisphere. As of 2005 it was calculated to be located at 79.74°S 108.22°E, near the Vostok Station. Because the field is not an exact dipole, the South Geomagnetic Pole does not coincide with the South Magnetic Pole. Furthermore, the South Geomagnetic Pole is wandering for the same reason its northern magnetic counterpart wanders.

Geomagnetic Reversal

Over the life of the Earth, the orientation of Earth's magnetic field has reversed many times, with geomagnetic north becoming geomagnetic south and vice versa – an event known as a geomagnetic reversal. Evidence of geomagnetic reversals can be seen at mid-ocean ridges where tectonic plates move apart. As magma seeps out of the mantle and solidifies to become new ocean floor, the magnetic minerals in it are magnetized in the direction of the magnetic field. Thus, starting at the most recently formed ocean floor, one can read out the direction of the magnetic field in previous times as one moves further away to older ocean floor.

A geomagnetic reversal is a change in a planet's magnetic field such that the positions of magnetic north and magnetic south are interchanged. The Earth's field has alternated between periods of *normal* polarity, in which the direction of the field was the same as the present direction, and *reverse* polarity, in which the field was the opposite. These periods are called *chrons*. The time spans of chrons are randomly distributed with most being between 0.1 and 1 million years with an average of 450,000 years. Most reversals are estimated to take between 1,000 and 10,000 years. The latest one, the Brunhes–Matuyama reversal, occurred 780,000 years ago.

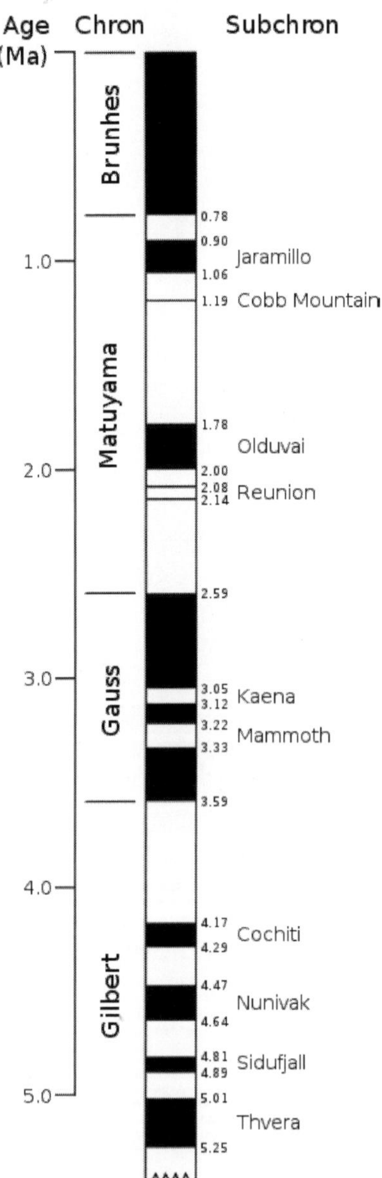

Figure: *Geomagnetic polarity during the last 5 million years (Pliocene and Quaternary, late Cenozoic Era). Dark areas denote periods where the polarity matches today's polarity, light areas denote periods where that polarity is reversed.*

A brief complete reversal, known as the Laschamp event, occurred only 41,000 years ago during the last glacial period. That reversal lasted only about 440 years with the actual change of polarity lasting around 250 years. During this change the strength of the magnetic field dropped to 5% of its present strength. Brief disruptions that do not result in reversal are called geomagnetic excursions.

History

In the early 20th century geologists first noticed that some volcanic rocks were magnetized opposite to the direction of the local Earth's field. The first estimate of the timing of magnetic reversals was made in the 1920s by Motonori Matuyama, who observed that rocks with reversed fields were all of early Pleistocene age or older. At the time, the Earth's polarity was poorly understood and the possibility of reversal aroused little interest.

Three decades later, when Earth's magnetic field was better understood, theories were advanced suggesting that the Earth's field might have reversed in the remote past. Most paleomagnetic research in the late 1950s included an examination of the wandering of the poles and continental drift. Although it was discovered that some rocks would reverse their magnetic field while cooling, it became apparent that most magnetized volcanic rocks preserved traces of the Earth's magnetic field at the time the rocks had cooled. In the absence of reliable methods for obtaining absolute ages for rocks, it was thought that reversals occurred approximately every million years.

The next major advance in understanding reversals came when techniques for radiometric dating were developed in the 1950s. Allan Cox and Richard Doell, at the United States Geological Survey, wanted to know whether reversals occurred at regular intervals, and invited the geochronologist Brent Dalrymple to join their group. They produced the first magnetic-polarity time scale in 1959. As they accumulated data, they continued to refine this scale in competition with Don Tarling and Ian McDougall at the Australian National University. A group led by Neil Opdyke at the Lamont-Doherty Geological Observatory showed that the same pattern of reversals was recorded in sediments from deep-sea cores.

During the 1950s and 1960s information about variations in the Earth's magnetic field was gathered largely by means of research vessels. But the complex routes of ocean cruises rendered the association of navigational data with magnetometre readings difficult. Only when data were plotted on a map did it become apparent that remarkably regular and continuous magnetic stripes appeared on the ocean floors.

In 1963 Frederick Vine and Drummond Matthews provided a simple explanation by combining the seafloor spreading theory of Harry Hess with the known time scale of reversals: if new sea floor is magnetized in the direction of the field, then it will change its polarity when the field reverses.

Thus, sea floor spreading from a central ridge will produce magnetic stripes parallel to the ridge. Canadian L. W. Morley independently proposed a similar explanation in January 1963, but his work was rejected by the scientific journals *Nature* and *Journal of Geophysical Research*, and remained unpublished until 1967, when it appeared in the literary magazine *Saturday Review*. The Morley–Vine–Matthews hypothesis was the first key scientific test of the seafloor spreading theory of continental drift.

Beginning in 1966, Lamont–Doherty Geological Observatory scientists found that the magnetic profiles across the Pacific-Antarctic Ridge were symmetrical and matched the pattern in the north Atlantic's Reykjanes ridges. The same magnetic anomalies were found over most of the world's oceans, which permitted estimates for when most of the oceanic crust had developed.

Observing Past Fields

Past field reversals can be and have been recorded in the "frozen" ferromagnetic (or more accurately, ferrimagnetic) minerals of consolidated sedimentary deposits or cooled volcanic flows on land.

The past record of geomagnetic reversals was first noticed by observing the magnetic stripe "anomalies" on the ocean floor. Lawrence W. Morley, Frederick John Vine and Drummond Hoyle Matthews made the connection to seafloor spreading in the Morley-Vine-Matthews hypothesis which soon led to the development of the theory of plate tectonics. The relatively constant rate at which the sea floor spreads results in substrate "stripes" from which past magnetic field polarity can be inferred from data gathered from towing a magnetometre along the sea floor.

Because no existing unsubducted sea floor (or sea floor thrust onto continental plates) is more than about 180 million years (Ma) old, other methods are necessary for detecting older reversals. Most sedimentary rocks incorporate tiny amounts of iron rich minerals, whose orientation is influenced by the ambient magnetic field at the time at which they formed. These rocks can preserve a record of the field if it is not later erased by chemical, physical or biological change.

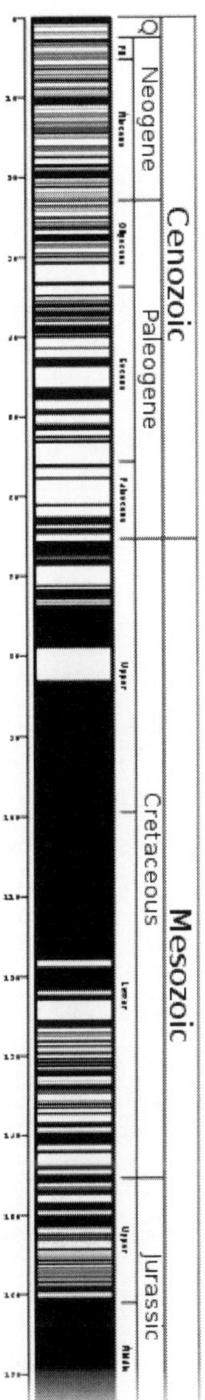

Figure: *Geomagnetic polarity since the middle Jurassic. Dark areas denote periods where the polarity matches today's polarity, light areas denote periods where that polarity is reversed.*

Because the magnetic field is global, similar patterns of magnetic variations at different sites may be used to correlate age in different locations. In the past four decades much paleomagnetic data about seafloor ages (up to ~250 Ma) has been collected and is useful in estimating the age of geologic sections. Not an independent dating method, it depends on "absolute" age dating methods like radioisotopic systems to derive numeric ages. It has become especially useful to metamorphic and igneous geologists where index fossils are seldom available.

Geomagnetic Polarity Time Scale

Through analysis of seafloor magnetic anomalies and dating of reversal sequences on land, paleomagnetists have been developing a *Geomagnetic Polarity Time Scale* (GPTS). The current time scale contains 184 polarity intervals in the last 83 million years.

Changing Frequency over Time

The rate of reversals in the Earth's magnetic field has varied widely over time. 72 million years ago (Ma), the field reversed 5 times in a million years. In a 4-million-year period centred on 54 Ma, there were 10 reversals; at around 42 Ma, 17 reversals took place in the span of 3 million years. In a period of 3 million years centring on 24 Ma, 13 reversals occurred. No fewer than 51 reversals occurred in a 12-million-year period, centring on 15 million years ago. Two reversals occurred during a span of 50,000 years. These eras of frequent reversals have been counterbalanced by a few "superchrons" – long periods when no reversals took place.

Superchrons

A *superchron* is a polarity interval lasting at least 10 million years. There are two well-established superchrons, the Cretaceous Normal and the Kiaman. A third candidate, the Moyero, is more controversial. The Jurassic Quiet Zone in ocean magnetic anomalies was once thought to represent a superchron, but is now attributed to other causes.

The *Cretaceous Normal* (also called the *Cretaceous Superchron* or C34) lasted for almost 40 million years, from about 120 to 83 million years ago, including stages of the Cretaceous period from the Aptian through the Santonian. The frequency of magnetic reversals steadily decreased prior to the period, reaching its low point (no reversals) during the period. Between the Cretaceous Normal and the present, the frequency has generally increased slowly.

The *Kiaman Reverse Superchron* lasted from approximately the late Carboniferous to the late Permian, or for more than 50 million years, from around 312 to 262 million years ago. The magnetic field had reversed polarity. The name "Kiaman" derives from the Australian village of Kiama, where some of the first geological evidence of the superchron was found in 1925.

The Ordovician is suspected to host another superchron, called the *Moyero Reverse Superchron,* lasting more than 20 million years (485 to 463 million years ago) . But until now this possible superchron has only been found in the Moyero river section north of the polar circle in Siberia. Moreover, the best data from elsewhere in the world do not show evidence for this superchron.

Certain regions of ocean floor, older than 160 Ma, have low-amplitude magnetic anomalies that are hard to interpret. They are found off the east coast of North America, the northwest coast of Africa, and the western Pacific. They were once thought to represent a superchron called the *Jurassic Quiet Zone,* but magnetic anomalies are found on land during this period. The geomagnetic field is known to have low intensity between about 130 Ma and 170 Ma, and these sections of ocean floor are especially deep, so the signal is attenuated between the floor and the surface.

Statistical Properties of Reversals

Several studies have analyzed the statistical properties of reversals in the hope of learning something about their underlying mechanism. The discriminating power of statistical tests is limited by the small number of polarity intervals. Nevertheless, some general features are well established. In particular, the pattern of reversals is random. There is no correlation between the lengths of polarity intervals. There is no preference for either normal or reversed polarity, and no statistical difference between the distributions of these polarities. This lack of bias is also a robust prediction of dynamo theory. Finally, as mentioned above, the rate of reversals changes over time.

The randomness of the reversals is inconsistent with periodicity, but several authors have claimed to find periodicity. However, these results are probably artifacts of an analysis using sliding windows to determine reversal rates.

Most statistical models of reversals have analyzed them in terms of a Poisson process or other kinds of renewal process. A Poisson process would have, on average, a constant reversal rate, so it is common to use a non-stationary Poisson process. However, compared to a Poisson process, there is a reduced probability of reversal for tens of thousands of years after a reversal. This could be due to an inhibition in the underlying mechanism, or it could just mean that some shorter polarity intervals have been missed. A random reversal pattern with inhibition can be represented by a gamma process. In 2006, a team of physicists at the University of Calabria found that the reversals also conform to a Lévy distribution, which describes stochastic processes with long-ranging correlations between events in time. The data are also consistent with a deterministic, but chaotic, process.

Character of Transitions

Duration: Most estimates for the duration of a polarity transition are between 1,000 and 10,000 years. However, studies of 15 million year old lava flows on Steens Mountain, Oregon, indicate that the Earth's magnetic field is capable of shifting at a rate of up to 6 degrees per day. This was initially met with skepticism from paleomagnetists. Even if changes occur that quickly in the core, the mantle, which is a semiconductor, is thought to act as a low-pass filter, removing variations with periods less than a few months.

A variety of possible rock magnetic mechanisms were proposed that would lead to a false signal. However, paleomagnetic studies of other sections from the same region (the Oregon Plateau flood basalts) give consistent results. It appears that the reversed-to-normal polarity transition that marks the end of Chron C5Cr (16.7 million years ago) contains a series of reversals and excursions. In addition, geologists Scott Bogue of Occidental College and Jonathan Glen of the US Geological Survey, sampling lava flows in Battle Mountain, Nevada, found evidence for a brief, several year long interval during a reversal when the field direction changed by over 50°. The reversal was dated to approximately 15 million years ago.

Magnetic Field

The magnetic field will not vanish completely, but many poles might form chaotically in different places during reversal, until it stabilizes again.

Causes

The magnetic field of the Earth, and of other planets that have magnetic fields, are generated by dynamo action in which convection of molten iron in the planetary core generates electric currents which in turn give rise to magnetic fields. In simulations of planetary dynamos, reversals often emerge spontaneously from the underlying dynamics. For example, Gary Glatzmaier and collaborator Paul Roberts of UCLA ran a numerical model of the coupling between electromagnetism and fluid dynamics in the Earth's interior. Their simulation reproduced key features of the magnetic field over more than 40,000 years of simulated time and the computer-generated field reversed itself. Global field reversals at irregular intervals have also been observed in the laboratory liquid metal experiment VKS2.

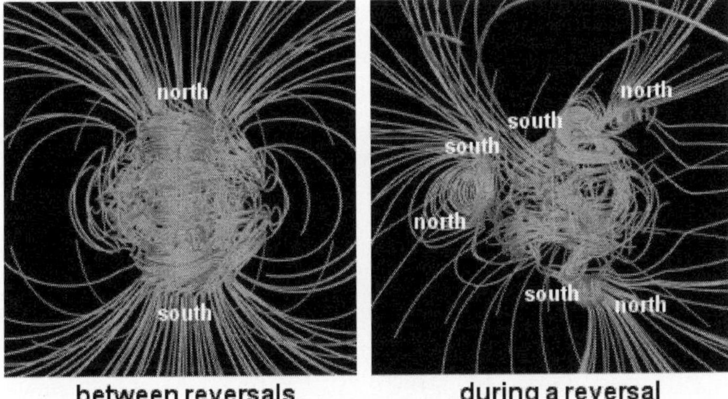

Figure: *NASA computer simulation using the model of Glatzmaier and Roberts. The tubes represent magnetic field lines, blue when the field points towards the centre and yellow when away. The rotation axis of the Earth is centred and vertical. The dense clusters of lines are within the Earth's core.*

In some simulations, this leads to an instability in which the magnetic field spontaneously flips over into the opposite orientation. This scenario is supported by observations of the solar magnetic field, which undergoes spontaneous reversals every 9–12 years. However, with the Sun it is observed that the solar magnetic intensity greatly increases during a reversal, whereas reversals on Earth seem to occur during periods of low field strength.

Hypothesized Triggers

Some scientists, such as Richard A. Muller, believe that geomagnetic reversals are not spontaneous processes but rather are triggered by external events that directly disrupt the flow in the Earth's core. Proposals include impact events or internal events such as the arrival of continental slabs carried down into the mantle by the action of plate tectonics at subduction zones or the initiation of new mantle plumes from the core-mantle boundary. Supporters of this theory hold that any of these events could lead to a large scale disruption of the dynamo, effectively turning off the geomagnetic field. Because the magnetic field is stable in either the present North-South orientation or a reversed orientation, they propose that when the field recovers from such a disruption it spontaneously chooses one state or the other, such that half the recoveries become reversals. However, the proposed mechanism does not appear to work in a quantitative model, and the evidence from stratigraphy for a correlation between reversals and impact events is weak. Most strikingly, there is no evidence for a reversal connected with the impact event that caused the Cretaceous–Paleogene extinction event.

Effects on Biosphere

Not long after the first geomagnetic polarity time scales were produced, scientists began exploring the possibility that reversals could be linked to extinctions. Most such proposals rest on the assumption that the Earth's magnetic field would be much weaker during reversals. Possibly the first such hypothesis was that high energy particles trapped in the Van Allen radiation belt could be liberated and bombard the Earth. Detailed calculations confirm that, if the Earth's dipole field disappeared entirely (leaving the quadrupole and higher components), most of the atmosphere would become accessible to high energy particles, but would act as a barrier to them, and cosmic ray collisions would produce secondary radiation of beryllium-10 or chlorine-36. An increase of beryllium-10 was noted in a 2012 German study showing a peak of beryllium-10 in Greenland ice cores during a brief complete reversal 41,000 years ago which led to the magnetic field strength dropping to an estimated 5% of normal during the reversal. There is evidence that this occurs both during secular variation and during reversals.

Another hypothesis by McCormac and Evans assumes that the Earth's field would disappear entirely during reversals. They argue that the atmosphere of Mars may have been eroded away by the solar wind because it had no magnetic field to protect it. They predict that ions would be stripped away from Earth's atmosphere above 100 km. However, the evidence from paleointensity measurements is that the magnetic field does not disappear. Based on paleointensity data for the last 800,000 years, the magnetopause is still estimated to be at about 3 Earth radii during the Brunhes-Matuyama reversal. Even if the magnetic field disappeared, the solar wind may induce a sufficient magnetic field in the Earth's ionosphere to shield the surface from energetic particles.

Hypotheses have also been advanced linking reversals to mass extinctions. Many such arguments were based on an apparent periodicity in the rate of reversals; more careful analyses show that the reversal record is not periodic. It may be, however, that the ends of superchrons have caused vigorous convection leading to widespread volcanism, and that the subsequent airborne ash caused extinctions.

Tests of correlations between extinctions and reversals are difficult for a number of reasons. Larger animals are too scarce in the fossil record for good statistics, so paleontologists have analyzed microfossil extinctions. Even microfossil data can be unreliable if there are hiatuses in the fossil record. It can appear that the extinction occurs at the end of a polarity interval when the rest of that polarity interval was simply eroded away. Statistical analysis shows no evidence for a correlation between reversals and extinctions.

Magnetic Anomaly

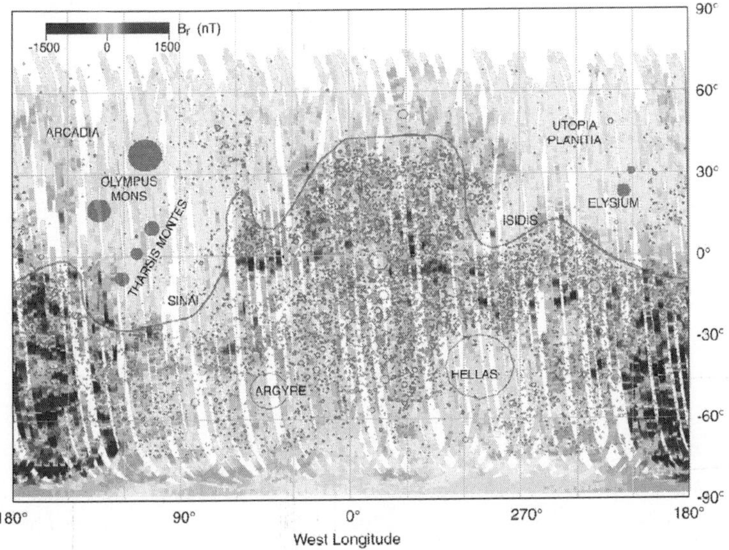

Figure: *Satellite measurements of the radial magnetic field on Mars.*

In geophysics, a magnetic anomaly is a local variation in the Earth's magnetic field resulting from variations in the chemistry or magnetism of the rocks. Mapping of variation over an area is valuable in detecting structures obscured by overlying material. The magnetic variation in successive bands of ocean floor parallel with mid-ocean ridges is important evidence supporting the theory of seafloor spreading, central to plate tectonics.

Measurement

Magnetic anomalies are generally a small fraction of the magnetic field. The total field ranges from 25,000 to 65,000 nanoteslas (nT). To measure anomalies, magnetometres need a sensitivity of 10 nT or less. There are three main types of magnetometre used to measure megnetic anomalies:

1. The fluxgate magnetometre was developed during World War II to detect submarines. It measures the component along a particular axis of the sensor, so it needs to be oriented. On land, it is often oriented vertically, while in aircraft, ships and satellites it is usually oriented so the axis is in the direction of the field. It measures the magnetic field continuously, but drifts

over time. One way to correct for drift is to take repeated measurements at the same place during the survey.

2. The proton precession magnetometre measures the strength of the field but not its direction, so it does not need to be oriented. Each measurement takes a second or more. It is used in most ground surveys except for boreholes and high-resolution gradiometre surveys.
3. Optically pumped magnetometres, which use alkali gases (most commonly rubidium and cesium) have high sample rates and sensitivities of 0.001 nT or less, but are more expensive than the other types of magnetometres. They are used on satellites and in most aeromagnetic surveys.

Data Acquisition

Ground-Based: In ground-based surveys, measurements are made at a series of stations, typically 15 to 60 m apart. Usually a proton precession magnetometre is used and it is often mounted on a pole. Raising the magnetometre reduces the influence of small ferrous objects that were discarded by humans. To further reduce unwanted signals, they do not carry objects such as keys, knives or compasses. In addition, objects such as motor vehicles, railway lines, and barbed wire fences are avoided. If some such contaminant is overlooked, it often shows up as a sharp spike in the anomaly, so such features are treated with suspicion. The main application for ground-based surveys is detailed search for minerals.

Aeromagnetic

Airborne magnetic surveys are often used in oil surveys to provide preliminary information for seismic surveys. In some countries such as Canada, government agencies have made systematic surveys of large areas. The survey generally involves making a series of parallel runs at a constant height and intervals of anywhere from a hundred metres to several kilometres. These are crossed by occasional tie lines, perpendicular to the main survey, to check for errors. The plane is a source of magnetism, so sensors are either mounted on a boom or towed behind on a cable. Aeromagnetic surveys have a lower spatial resolution of ground surveys, but this can be an advantage for a regional survey of deeper rocks.

Shipborne

In shipborne surveys, a magnetometre is towed a few hundred metres behind a ship in a device called a *fish*. The sensor is kept at a constant depth of about 15 m. Otherwise, the procedure is similar to that used in aeromagnetic surveys.

Spacecraft

American and Russian satellites began measuring the Earth's field in the 1960's. In the fall of 1979, Magsat was launched and jointly operated by NASA and USGS until the spring of 1980. It had a cesium vapor scalar magnetometre and a fluxgate vector magnetometre. CHAMP, a German satellite, made precise gravity and magnetic measurements from 2001 to 2010. A Danish satellite, Ørsted, was launched in 1999 and is still in operation, while the Swarm mission of the European Space Agency involves a "constellation" of three satellites that were launched in November, 2013.

Data Reduction

There are two main corrections that are needed for magnetic measurements. The first is remove short-term variations in the field from external sources. There are *diurnal variations* that have a period of 24 hours and magnitudes of up to 30 nT, probably from the action of the solar wind on the ionosphere. In addition, magnetic storms can have peak magnitudes of 1000 nT and can last for several days. Their contribution can be measured by returning to a base station repeatedly or by having another magnetometre that periodically measures the field at a fixed location.

The anomaly is the local contribution to the magnetic field, so the main geomagnetic field must be subtracted from it. Usually the International Geomagnetic Reference Field is used for this purpose. This is a large-scale, time-averaged mathematical model of the Earth's field based on measurements from satellites, magnetic observatories and other surveys.

Some corrections that are needed for gravity anomalies are less important for magnetic anomalies. For example, the vertical gradient of the magnetic field is 0.03 nT/m or less, so an elevation correction is generally not needed.

Interpretation

The magnetization in the surveyed rock is a vector sum of induced and remanent magnetization:

$$M = M_i + M_r.$$

The induced magnetization of many minerals is the product of the ambient magnetic field and their magnetic susceptibility ÷:

$$M_i = \chi H.$$

Some susceptibilities are given in the table.

Magnetic susceptibilities of common rocks and minerals	
Type	Susceptibility ($\times 10^{-3}$ SI)

Minerals that are diamagnetic or paramagnetic only have an induced magnetization. Ferromagnetic minerals such as magnetite also can carry a remanent magnetization or remanence. This remanence can last for millions of years, so it may be in a completely different direction from the present Earth's field. If a remanence is present, it is difficult to separate from the induced magnetization unless samples of the rock are measured. The ratio of the magnitudes, $Q = M_r/M_i$, is called the Koenigsberger ratio.

Applications

Ocean Floor Stripes

Magnetic surveys over the oceans have revealed a characteristic pattern of anomalies around mid-ocean ridges. They involve a series of positive and negative anomalies in the intensity of the magnetic field, forming stripes running parallel to each ridge. They are often symmetric about the axis of the ridge. The stripes are generally tens of kilometres wide, and the anomalies are a few hundred nanoteslas. The source of these anomalies is primarily permanent magnetization carried by titanomagnetite minerals in basalt and gabbros. They are magnetized when ocean crust is formed at the ridge.

As magma rises to the surface and cools, the rock acquires a thermoremanent magnetization in the direction of the field. Then the rock is carried away from the ridge by the motions of the tectonic plates. Every few hundred thousand years, the direction of the magnetic field reverses. Thus, the pattern of stripes is a global phenomenon and can be used to calculate the velocity of seafloor spreading.

Geomagnetic Storm

A geomagnetic storm is a temporary disturbance of the Earth's magnetosphere caused by a solar wind shock wave and/or cloud of magnetic field which interacts with the Earth's magnetic field. The increase in the solar wind pressure initially compresses the magnetosphere and the solar wind's magnetic field interacts with the Earth's magnetic field and transfers an increased energy into the magnetosphere. Both interactions cause an increase in movement of plasma through the magnetosphere (driven by increased electric fields inside the magnetosphere) and an increase in electric current in the magnetosphere and ionosphere.

Figure: *Artist's depiction of solar wind particles interacting with Earth's magnetosphere. Sizes are not to scale.*

During the main phase of a geomagnetic storm, electric current in the magnetosphere creates a magnetic force which pushes out the boundary between the magnetosphere and the solar wind.

The disturbance in the interplanetary medium which drives the geomagnetic storm may be due to a solar coronal mass ejection (CME) or a high speed stream (co-rotating interaction region or CIR) of the solar wind originating from a region of weak magnetic field on the Sun's surface. The frequency of geomagnetic storms increases and decreases with the sunspot cycle. CME driven storms are more common during the maximum of the solar cycle and CIR driven storms are more common during the minimum of the solar cycle.

There are several space weather phenomena which tend to be associated with or are caused by a geomagnetic storm.

These include: Solar Energetic Particle (SEP) events, geomagnetically induced currents (GIC), ionospheric disturbances which cause radio and radar scintillation, disruption of navigation by magnetic compass and auroral displays at much lower latitudes than normal.

In 1989, a geomagnetic storm energized ground induced currents which disrupted electric power distribution throughout most of the province of Quebec and caused aurorae as far south as Texas.

Definition of a Geomagnetic Storm

A geomagnetic storm is defined by changes in the DST (disturbance – storm time) index. The Dst index estimates the globally averaged change of the horizontal component of the Earth's magnetic field at the magnetic equator based on measurements from a few magnetometre stations. Dst is computed once per hour and reported in near-real-time. During quiet times, Dst is between +20 and -20 nano-Tesla (nT).

A geomagnetic storm has three phases: an initial phase, a main phase and a recovery phase. The initial phase is characterized by Dst (or its one-minute component SYM-H) increasing by 20 to 50 nT in tens of minutes. The initial phase is also referred to as a storm sudden commencement (SSC). However, not all geomagnetic storms have an initial phase and not all sudden increases in Dst or SYM-H are followed by a geomagnetic storm.

The main phase of a geomagnetic storm is defined by Dst decreasing to less than -50 nT. The selection of -50 nT to define a storm is somewhat arbitrary. The minimum value during a storm will be between -50 and approximately -600 nT. The duration of the main phase is typically between 2 and 8 hours. The recovery phase is the period when Dst changes from its minimum value to its quiet time value. The period of the recovery phase may be as short as 8 hours or as long as 7 days.

The size of a geomagnetic storm is classified as moderate (-50 nT > minimum of Dst < -100 nT), intense (-100 nT > minimum Dst < -250 nT) or super-storm (minimum of Dst > -250 nT).

Historical Occurrences

The first observation of the effects of a geomagnetic storm occurred early in the 19th century: From May 1806 until June 1807 the German Alexander von Humboldt recorded the bearing of a magnetic compass in Berlin. On 21 December 1806 he noticed that his compass had become erratic during a bright auroral event.

On September 1 – 2, 1859, the largest recorded geomagnetic storm occurred. From August 28 until September 2, 1859, numerous sunspots and solar flares were observed on the Sun, the largest flare occurring on September 1.

This is referred to as the Solar storm of 1859 or the Carrington Event. It can be assumed that a massive coronal mass ejection (CME), associated with the flare, was launched from the Sun and reached the Earth within eighteen hours — a trip that normally takes three to four days.

The horizontal intensity of geomagnetic field was reduced by 1600 nT as recorded by the Colaba Observatory. It is estimated that Dst would have been approximately -1760 nT. Telegraph wires in both the United States and Europe experienced induced emf, in some cases even shocking telegraph operators and causing fires. Aurorae were seen as far south as Hawaii, Mexico, Cuba, and Italy — phenomena that are usually only seen near the poles. Ice cores show evidence that events of similar intensity recur at an average rate of approximately once per 500 years.

Since 1859, less severe storms have occurred, notably the aurora of November 17, 1882 and the May 1921 geomagnetic storm, both with disruption of telegraph service and initiation of fires, and 1960, when widespread radio disruption was reported.

The March 1989 geomagnetic storm caused the collapse of the Hydro-Québec power grid in a matter of seconds as equipment protection relays tripped in a cascading sequence of events. Six million people were left without power for nine hours, with significant economic loss. The storm even caused aurorae as far south as Texas. The geomagnetic storm causing this event was itself the result of a coronal mass ejection, ejected from the Sun on March 9, 1989. The minimum of Dst was -589 nT.

On July 14, 2000, an X5 class flare erupted on the Sun (known as the Bastille Day event) and a coronal mass ejection was launched directly at the Earth. A geomagnetic super storm occurred on July 15–17; the minimum of the Dst index was – 301 nT. Despite the strength of the geomagnetic storm, no electrical power distribution failures were reported. The Bastille Day event was observed by Voyager 1 and Voyager 2, thus it is the farthest out in the solar system that a solar storm has been observed.

Seventeen major flares erupted on the Sun between 19 October and 5 November 2003, including perhaps the most intense flare ever measured on the GOES XRS sensor – a huge X28 flare, resulting in an extreme radio blackout, on 4 November. These flares were associated with CME events which impacted the Earth. The CMEs caused three geomagnetic storms between Oct 29 and November 2 during which the second and third storms were initiated before the previous storm period had fully recovered. The minimum Dst values were -151, -353 and -383 nT. Another storm in this event period occurred on November 4 – 5 with a minimum Dst of -69.nT. The last geomagnetic storm was weaker than the preceding storms because the active region on the Sun had rotated beyond the meridian where the central portion CME created during the flare event passed to the side of the Earth. The whole sequence of events is known as the Halloween Solar Storm. The Wide Area Augmentation System (WAAS) operated by the Federal Aviation Administration (FAA) was offline for approximately 30 hours due to the storm. The Japanese ADEOS-2 satellite was severely damaged and the operation of many other satellites were interrupted due to the storm.

Interactions with Planetary Processes

The solar wind also carries with it the magnetic field of the Sun. This field will have either a North or South orientation. If the solar wind has energetic bursts, contracting and expanding the magnetosphere, or if the solar wind takes a southward polarization, geomagnetic storms can be expected. The southward field causes magnetic reconnection of the dayside magnetopause, rapidly injecting magnetic and particle energy into the Earth's magnetosphere. During a geomagnetic storm, the ionosphere's F_2 layer will become unstable, fragment, and may even disappear. In the northern and southern pole regions of the Earth, auroras will be observable in the sky.

Instruments for Researching Geomagnetic Storms

Magnetometres monitor the auroral zone as well as the equatorial region. Two types of radar — coherent scatter and incoherent scatter — are used to probe the auroral ionosphere. By bouncing signals off ionospheric irregularities (which convect with their field lines) one can trace their motion and infer magnetospheric convection.

Spacecraft Instruments Include:

- Magnetometres, usually of the flux gate type. Usually these are at the end of booms, to keep them away from magnetic interference by the spacecraft and its electric circuits.
- Electric sensors at the ends of opposing booms are used to measure potential differences between separated points, to derive electric field associated with convection. The method works best at high plasma densities in low Earth orbit; far from Earth long booms are needed, to avoid shielding-out of electric forces.
- Radio sounders from the ground can bounce radio waves of varying frequency off the ionosphere, and by timing their return get the profile of electron density in the ionosphere — up to its peak, past which radio waves no longer return. Radio sounders in low Earth orbit aboard the Canadian Alouette 1 (1962) and

Alouette 2 (1965), beamed radio waves earthward and observed the electron density profile of the "topside ionosphere." Other radio sounding methods were also tried in the ionosphere (e.g. on IMAGE).

- A great variety of "particle detectors" have operated in orbit. The original observations of the Van Allen radiation belt used a Geiger counter, a crude detector unable to tell particle charge or energy. Later scintillator detectors were used, and still later "channeltron" electron multipliers have found particularly wide use. To derive charge and mass composition, as well as energies, a variety of mass spectrograph designs were used. For energies up to about 50 keV (which constitute most of the magnetospheric plasma) time-of-flight spectrometres (e.g. "top-hat" design) are widely used.

Computers have made it possible to bring together decades of isolated magnetic observations and extract average patterns of electrical currents and average responses to interplanetary variations. They also run simulations of the global magnetosphere and its responses, by solving the equations of magnetohydrodynamics (MHD) on a numerical grid. Appropriate extensions must be added to cover the inner magnetosphere, where magnetic drifts and ionospheric conduction also need to be taken into account. So far the results are difficult to interpret, and certain assumptions are still needed to cover small-scale phenomena.

Geomagnetic Storm Effects

Radiation Hazards to Humans: Intense solar flares release very-high-energy particles that can cause radiation poisoning to humans (and mammals in general) in the same way as low-energy radiation from nuclear blasts.

Earth's atmosphere and magnetosphere allow adequate protection at ground level, but astronauts in space are subject to potentially lethal doses of radiation. The penetration of high-energy particles into living cells can cause chromosome damage, cancer, and a host of other health problems. Large doses can be fatal immediately.

Solar protons with energies greater than 30 MeV are particularly hazardous. In October 1989, the Sun produced enough energetic particles that, if an astronaut were to have been standing on the Moon at the time, wearing only a space suit and caught out in the brunt of the storm, they would probably have died; the expected dose would be about 7000 rem. Note that astronauts who had time to gain safety in a shelter beneath moon soil would have absorbed only slight amounts of radiation. Solar proton events can also produce elevated radiation aboard aircraft flying at high altitudes. Although these risks are small, monitoring of solar proton events by satellite instrumentation allows the occasional exposure to be monitored and evaluated, and eventually the flight paths and altitudes adjusted in order to lower the absorbed dose of the flight crews.

Fauna and Flora

Possibly the most closely studied of the variable Sun's biological effects has been the degradation of homing pigeons' navigational abilities during geomagnetic storms. Pigeons and other migratory animals, such as dolphins and whales, demonstrate magnetosensitive behavioural responses that were once thought to be mediated by neurons which contained the mineral magnetite located in the beak. However, the basis of sensory perception of magnetic fields is currently unknown.

Disruption of Electrical Systems

Odenwald suggests that a geomagnetic storm on the scale of the solar storm of 1859 today would cause billions of dollars of damage to satellites, power grids and radio communications, and could cause electrical blackouts on a massive scale that might not be repaired for weeks.

Communications

Many communication systems use the ionosphere to reflect radio signals over long distances. Ionospheric storms can affect radio communication at all latitudes. Some radio frequencies are absorbed and others are reflected, leading to rapidly fluctuating signals and unexpected propagation paths. TV and commercial radio stations are little affected by solar activity, but ground-to-air, ship-to-shore, shortwave broadcast, and amateur radio (mostly the bands below 30 MHz) are frequently disrupted. Radio operators using HF bands rely upon solar and geomagnetic alerts to keep their communication circuits up and running.

Some military detection or early warning systems are also affected by solar activity. The *over-the-horizon radar* bounces signals off the ionosphere to monitor the launch of aircraft and missiles from long distances. During geomagnetic storms, this system can be severely hampered by radio clutter. Some submarine detection systems use the magnetic signatures of submarines as one input to their locating schemes. Geomagnetic storms can mask and distort these signals.

The Federal Aviation Administration routinely receives alerts of solar radio bursts so that they can recognise communication problems and avoid unnecessary maintenance. When an aircraft and a ground station are aligned with the Sun, jamming of air-control radio frequencies can occur. This can also happen when an Earth station, a satellite, and the Sun are in alignment. In order to prevent unnecessary maintenance on satellite communications systems aboard aircraft AirSatOne provides a live feed for Geophysical Events from NOAA's Space Weather Prediction Centre. AirSatOne's live feed allows users to view observed and predicted space storms. Geophysical Alerts are important to flight crews and maintenance personnel to determine if any upcoming activity or history has or will have an effect on satellite communications, GPS navigation and HF Communications.

The telegraph lines in the past were affected by geomagnetic storms as well. The telegraphs used a single long wire for the data line, stretching for many miles, using ground as the return wire and being fed with DC power from a battery; this made them (together with the power lines mentioned below) susceptible to being influenced by the fluctuations caused by the ring current. The voltage/current induced by the geomagnetic storm could have led to diminishing of the signal, when subtracted from the battery polarity, or to overly strong and spurious signals when added to it; some operators in such cases even learned to disconnect the battery and rely on the induced current as their power source. In extreme cases the induced current was so high the coils at the receiving side burst in flames, or the operators received electric shocks. Geomagnetic storms affect also long-haul telephone lines, including undersea cables unless they are fiber optic.

Damage to communications satellites can disrupt non-terrestrial telephone, television, radio, and Internet links. The National Academy of Sciences reported in 2008 on possible scenarios of widespread disruption in the 2012–2013 solar peak.

Navigation Systems

Systems such as GPS, LORAN, and the now-defunct OMEGA are adversely affected when solar activity disrupts their signal propagation. The OMEGA system consisted of eight transmitters located throughout the world. Airplanes and ships used the very low frequency signals from these transmitters to determine their positions. During solar events and geomagnetic storms, the system gave navigators information that is inaccurate by as much as several miles. If navigators had been alerted that a proton event or geomagnetic storm was in progress, they could have switched to a backup system.

GPS signals are affected when solar activity causes sudden variations in the density of the ionosphere, causing the GPS signals to scintillate (like a twinkling star). The scintillation of satellite signals during ionospheric disturbances is studied at HAARP during ionospheric modification experiments. It has also been studied at the Jicamarca Radio Observatory.

One technology used to allow GPS receivers to continue to operate in the presence of some confusing signals is Receiver Autonomous Integrity Monitoring (RAIM). However, RAIM is predicated on the assumption that a majority of the GPS constellation is operating properly, and so it is much less useful when the entire constellation is perturbed by global influences such as geomagnetic storms. Even if RAIM detects a loss of integrity in these cases, it may not be able to provide a useful, reliable signal.

Satellite Hardware Damage

Geomagnetic storms and increased solar ultraviolet emission heat Earth's upper atmosphere, causing it to expand. The heated air rises, and the density at the orbit of satellites up to about 1,000 km (621 mi) increases significantly. This results in increased drag on satellites in space, causing them to slow and change orbit slightly. Unless Low Earth Orbit satellites are routinely boosted to higher orbits, they slowly fall, and eventually burn up in Earth's atmosphere.

Skylab is an example of a spacecraft re-entering Earth's atmosphere prematurely in 1979 as a result of higher-than-expected solar activity. During the great geomagnetic storm of March 1989, four of the Navy's navigational satellites had to be taken out of service for up to a week, the U.S. Space Command had to post new orbital elements for over 1000 objects affected, and the Solar Maximum Mission satellite fell out of orbit in December the same year.

The vulnerability of the satellites depends on their position as well. The South Atlantic Anomaly is a perilous place for a satellite to pass through.

As technology has allowed spacecraft components to become smaller, their miniaturized systems have become increasingly vulnerable to the more energetic solar particles. These particles can cause physical damage to microchips and can change software commands in satellite-borne computers.

Another problem for satellite operators is differential charging. During geomagnetic storms, the number and energy of electrons and ions increase. When a satellite travels through this energized environment, the charged particles striking the spacecraft cause different portions of the spacecraft to be differentially charged. Eventually, electrical discharges can arc across spacecraft components, harming and possibly disabling them.

Bulk charging (also called deep charging) occurs when energetic particles, primarily electrons, penetrate the outer covering of a satellite and deposit their charge in its internal parts. If sufficient charge accumulates in any one component, it may attempt to neutralize by discharging to other components. This discharge is potentially hazardous to the satellite's electronic systems.

Mains Electricity Grid

When magnetic fields move about in the vicinity of a conductor such as a wire, a geomagnetically induced current is produced in the conductor. This happens on a grand scale during geomagnetic storms (the same mechanism also influenced telephone and telegraph lines before fiber optics.

Long transmission lines (many kilometres in length) are thus subject to damage by this effect. Notably, this chiefly includes operators in China, North America, and Australia, especially in more modern high-voltage, low-resistance lines. The European grid consists mainly of shorter transmission circuits, which are less vulnerable to damage.

The (nearly direct) currents induced in these lines from geomagnetic storms are harmful to electrical transmission equipment, especially transformers — inducing core saturation, constraining their performance (as well as tripping various safety devices), and causing coils and cores to heat up.

In extreme cases, this heat can disable or destroy them, even inducing a chain reaction that can overload transformers throughout a system. Most generators are connected to the grid via transformers, isolating them from the induced currents on the grid, making them much less susceptible to damage due to geomagnetically induced current. However, a transformer that is subjected to this will act as an unbalanced load to the generator, causing negative sequence current in the stator and consequently heating of the rotor.

According to a study by Metatech corporation, a storm with a strength comparable to that of 1921 would destroy more than 300 transformers and leave over 130 million people without power, with a cost totaling several trillion dollars. A massive solar flare could knock out electric power for months.

These predictions are contradicted by a NERC report that concludes that a geomagnetic storm would cause temporary grid instability but no widespread destruction of high-voltage transformers. The report points out that the widely quoted Quebec grid collapse was not caused by overheating transformers but by the near-simultaneous tripping of seven relays.

By receiving geomagnetic storm alerts and warnings (e.g. by the Space Weather prediction Centre; via Space Weather satellites as SOHO or ACE), power companies can minimize damage to power transmission equipment, by momentarily disconnecting transformers or by inducing temporary blackouts. Preventative measures also exist, including preventing the inflow of GICs into the grid through the neutral-to-ground connection.

Geologic Exploration

Earth's magnetic field is used by geologists to determine subterranean rock structures. For the most part, these geodetic surveyors are searching for oil, gas, or mineral deposits. They can accomplish this only when Earth's field is quiet, so that true magnetic signatures can be detected.

Other geophysicists prefer to work during geomagnetic storms, when strong variations in the Earth's normal subsurface electric currents allow them to sense subsurface oil or mineral structures. This technique is called magnetotellurics. For these reasons, many surveyors use geomagnetic alerts and predictions to schedule their mapping activities.

Pipelines

Rapidly fluctuating geomagnetic fields can produce geomagnetically induced currents in pipelines. This can cause multiple problems for pipeline engineers. Flow metres in the pipeline can transmit erroneous flow information, and the corrosion rate of the pipeline is dramatically increased. If engineers incorrectly attempt to balance the current during a geomagnetic storm, corrosion rates may increase even more. Once again, pipeline managers thus receive space weather alerts and warnings to allow them to implement defencive measures.

Preparations against Solar Storms

Although geomagnetic storms that are severe enough to cause damage to the mains electricity grid are rare, there are some people who actively prepare against such events.

To eliminate the possibility of damage to private electrical equipment, houses can be protected with a (double-pole) circuit breaker on the power line that enters the house Since long powerlines (such as the ones coming from the mains electricity grid) gather and convey a huge amount of electric power from the solar storms, it is very likely that any connected domestic equipment that is connected to it will be damaged during a powerful solar storm.

Another technique often used is to place essential electrical equipment (such as electricity generators) in a cardboard box lined with aluminum on the outside. This effectively makes a Faraday cage, hence shielding the electrical equipment from any radiation. However, for radiation damage to occur, the geomagnetic storms would need to be very powerful, and/or the equipment would be of a type that is very vulnerable to radiation damage. Also, simply placing the equipment inside your house will already provide some protection against geomagnetic radiation.

Chapter 8

Tide

Tides are the rise and fall of sea levels caused by the combined effects of the gravitational forces exerted by the Moon and the Sun and the rotation of the Earth.

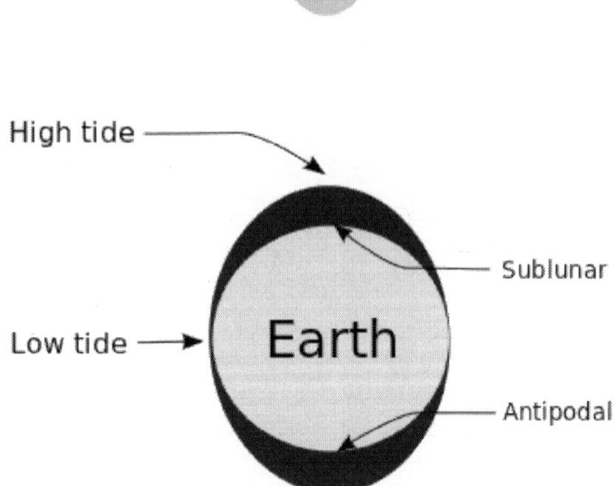

Figure: *Schematic of the lunar portion of earth's tides showing (exaggerated) high tides at the sublunar and antipodal points for the hypothetical case of an ocean of constant depth with no land. There would also be smaller, superimposed bulges on the sides facing toward and away from the sun.*

Some shorelines experience two almost equal high tides and two low tides each day, called a semi-diurnal tide. Some locations experience only one high and one low tide each day, called a diurnal tide. Some locations experience two uneven tides a day, or sometimes one high and one low each day; this is called a mixed tide.

The times and amplitude of the tides at a locale are influenced by the alignment of the Sun and Moon, by the pattern of tides in the deep ocean, by the amphidromic systems of the oceans, and by the shape of the coastline and near-shore bathymetry.

Tides vary on timescales ranging from hours to years due to numerous influences. To make accurate records, tide gauges at fixed stations measure the water level over time. Gauges ignore variations caused by waves with periods shorter than minutes.

These data are compared to the reference (or datum) level usually called mean sea level.

While tides are usually the largest source of short-term sea-level fluctuations, sea levels are also subject to forces such as wind and barometric pressure changes, resulting in storm surges, especially in shallow seas and near coasts.

Tidal phenomena are not limited to the oceans, but can occur in other systems whenever a gravitational field that varies in time and space is present. For example, the solid part of the Earth is affected by tides, though this is not as easily seen as the water tidal movements.

Characteristics

Tide changes proceed via the following stages:

- Sea level rises over several hours, covering the intertidal zone; flood tide.
- The water rises to its highest level, reaching high tide.
- Sea level falls over several hours, revealing the intertidal zone; ebb tide.
- The water stops falling, reaching low tide.

Tides produce oscillating currents known as tidal streams. The moment that the tidal current ceases is called slack water or slack tide. The tide then reverses direction and is said to be turning. Slack water usually occurs near high water and low water. But there are locations where the moments of slack tide differ significantly from those of high and low water.

Tides are commonly *semi-diurnal* (two high waters and two low waters each day), or *diurnal* (one tidal cycle per day). The two high waters on a given day are typically not the same height (the daily inequality); these are the *higher high water* and the *lower high water* in tide tables. Similarly, the two low waters each day are the *higher low water* and the *lower low water.* The daily inequality is not consistent and is generally small when the Moon is over the equator.

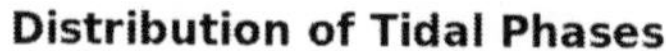

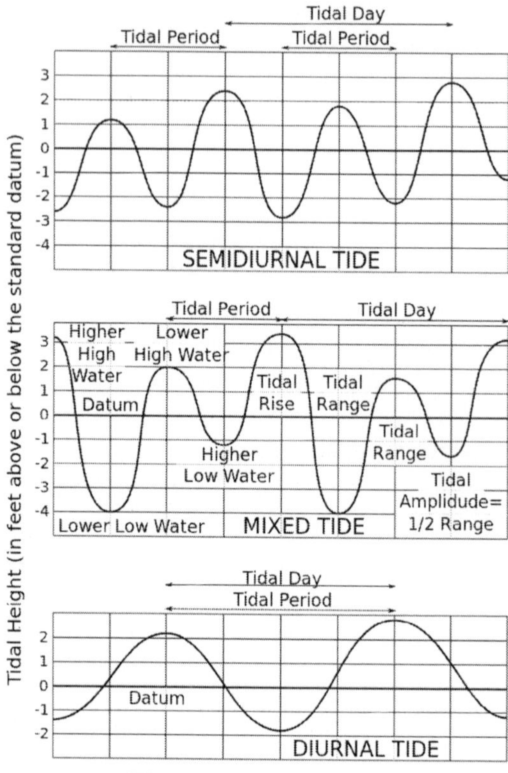

Figure: *Types of tides*

Tidal Constituents

Tidal changes are the net result of multiple influences that act over varying periods. These influences are called tidal constituents. The primary constituents are the Earth's rotation, the positions of the Moon and the Sun relative to Earth, the Moon's altitude (elevation) above the Earth's equator, and bathymetry.

Variations with periods of less than half a day are called *harmonic constituents*. Conversely, cycles of days, months, or years are referred to as *long period* constituents.

The tidal forces affect the entire earth, but the movement of the solid Earth is only centimetres. The atmosphere is much more fluid and compressible so its surface moves kilometres, in the sense of the contour level of a particular low pressure in the outer atmosphere.

Principal Lunar Semi-Diurnal Constituent

In most locations, the largest constituent is the "principal lunar semi-diurnal", also known as the *M2* (or M_2) tidal constituent. Its period is about 12 hours and 25.2 minutes, exactly half a *tidal lunar day*, which is the average time separating one lunar zenith from the next, and thus is the time required for the Earth to rotate once relative to the Moon. Simple tide clocks track this constituent. The lunar day is longer than the Earth day because the Moon orbits in the same direction the Earth spins. This is analogous to the minute hand on a watch crossing the hour hand at 12:00 and then again at about 1:05½ (not at 1:00).

The Moon orbits the Earth in the same direction as the Earth rotates on its axis, so it takes slightly more than a day—about 24 hours and 50 minutes—for the Moon to return to the same location in the sky. During this time, it has passed overhead (culmination) once and underfoot once (at an hour angle of 00:00 and 12:00 respectively), so in many places the period of strongest tidal forcing is the above-mentioned, about 12 hours and 25 minutes. The moment of highest tide is not necessarily when the Moon is nearest to zenith or nadir, but the period of the forcing still determines the time between high tides.

Because the gravitational field created by the Moon weakens with distance from the Moon, it exerts a slightly stronger than average force on the side of the Earth facing the Moon, and a slightly weaker force on the opposite side. The Moon thus tends to "stretch" the Earth slightly along the line connecting the two bodies.

The solid Earth deforms a bit, but ocean water, being fluid, is free to move much more in response to the tidal force, particularly horizontally. As the Earth rotates, the magnitude and direction of the tidal force at any particular point on the Earth's surface change constantly; although the ocean never reaches equilibrium—there is never time for the fluid to "catch up" to the state it would eventually reach if the tidal force were constant—the changing tidal force nonetheless causes rhythmic changes in sea surface height.

Semi-Diurnal Range Differences

When there are two high tides each day with different heights (and two low tides also of different heights), the pattern is called a *mixed semi-diurnal tide.*

Range Variation: Springs and Neaps

The semi-diurnal range (the difference in height between high and low waters over about half a day) varies in a two-week cycle. Approximately twice a month, around new moon and full moon when the Sun, Moon, and Earth form a line (a condition known as syzygy), the tidal force due to the sun reinforces that due to the Moon. The tide's range is then at its maximum; this is called the *spring tide.* It is not named after the season, but, like that word, derives from the meaning "jump, burst forth, rise", as in a natural spring.

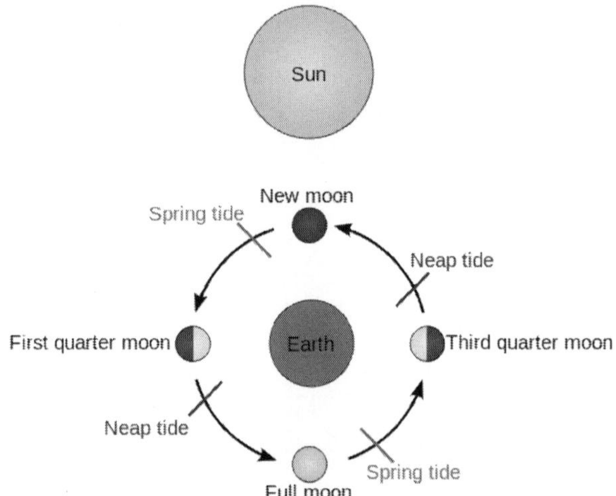

Figure: *The types of tides*

When the Moon is at first quarter or third quarter, the sun and Moon are separated by 90° when viewed from the Earth, and the solar tidal force partially cancels the Moon's. At these points in the lunar cycle, the tide's range is at its minimum; this is called the *neap tide*, or *neaps* (a word of uncertain origin).

Spring tides result in high waters that are higher than average, low waters that are lower than average, 'slack water' time that is shorter than average, and stronger tidal currents than average. Neaps result in less-extreme tidal conditions. There is about a seven-day interval between springs and neaps.

Lunar Altitude

The changing distance separating the Moon and Earth also affects tide heights. When the Moon is closest, at perigee, the range increases, and when it is at apogee, the range shrinks. Every 7½ lunations (the full cycles from full moon to new to full), perigee coincides with either a new or full moon causing perigean spring tides with the largest *tidal range*. Even at its most powerful this force is still weak causing tidal differences of inches at most.

Bathymetry

The shape of the shoreline and the ocean floor changes the way that tides propagate, so there is no simple, general rule that predicts the time of high water from the Moon's position in the sky. Coastal characteristics such as underwater bathymetry and coastline shape mean that individual location characteristics affect tide forecasting; actual high water time and height may differ from model predictions due to the coastal morphology's effects on tidal flow. However, for a given location the relationship between lunar altitude and the time of high or low tide (the lunitidal interval) is relatively constant and predictable, as is the time of high or low tide relative to other points on the same coast. For example, the high tide at Norfolk, Virginia, predictably occurs approximately two and a half hours before the Moon passes directly overhead.

Land masses and ocean basins act as barriers against water moving freely around the globe, and their varied shapes and sizes affect the size of tidal frequencies. As a result, tidal patterns vary. For example, in the U.S., the East coast has predominantly semi-diurnal tides, as do Europe's Atlantic coasts, while the West coast predominantly has mixed tides.

Other Constituents

These include solar gravitational effects, the obliquity (tilt) of the Earth's equator and rotational axis, the inclination of the plane of the lunar orbit and the elliptical shape of the Earth's orbit of the sun.

A compound tide (or overtide) results from the shallow-water interaction of its two parent waves.

Phase and Amplitude

Because the M_2 tidal constituent dominates in most locations, the stage or *phase* of a tide, denoted by the time in hours after high water, is a useful concept. Tidal stage is also measured in degrees, with 360° per tidal cycle. Lines of constant tidal phase are called *cotidal lines*, which are analogous to contour lines of constant altitude on topographical maps. High water is reached simultaneously along the cotidal lines extending from the coast out into the ocean, and cotidal lines (and hence tidal phases) advance along the coast. Semi-diurnal and long phase constituents are measured from high water, diurnal from maximum flood tide. This and the discussion that follows is precisely true only for a single tidal constituent.

For an ocean in the shape of a circular basin enclosed by a coastline, the *cotidal lines* point radially inward and must eventually meet at a common point, the amphidromic point. The amphidromic point is at once cotidal with high and low waters, which is satisfied by *zero* tidal motion. (The rare exception occurs when the tide encircles an island, as it does around New Zealand, Iceland and Madagascar.) Tidal motion generally lessens moving away from continental coasts, so that crossing the cotidal lines are contours of constant *amplitude* (half the distance between high and low water) which decrease to zero at the amphidromic point. For a semi-diurnal tide the amphidromic point can be thought of roughly like the centre of a clock face, with the hour hand pointing in the direction of the high water cotidal line, which is directly opposite the low water cotidal line.

High water rotates about the amphidromic point once every 12 hours in the direction of rising cotidal lines, and away from ebbing cotidal lines. This rotation is generally clockwise in the southern hemisphere and counterclockwise in the northern hemisphere, and is caused by the Coriolis effect.

The difference of cotidal phase from the phase of a reference tide is the *epoch*. The reference tide is the hypothetical constituent "equilibrium tide" on a landless Earth measured at 0° longitude, the Greenwich meridian.

In the North Atlantic, because the cotidal lines circulate counterclockwise around the amphidromic point, the high tide passes New York Harbor approximately an hour ahead of Norfolk Harbor. South of Cape Hatteras the tidal forces are more complex, and cannot be predicted reliably based on the North Atlantic cotidal lines.

Forces

The tidal force produced by a massive object (Moon, hereafter) on a small particle located on or in an extensive body (Earth, hereafter) is the vector difference between the gravitational force exerted by the Moon on the particle, and the gravitational force that would be exerted on the particle if it were located at the Earth's centre of mass. The solar *gravitational force* on the Earth is on average 179 times stronger than the lunar, but because the Sun is on average 389 times farther from the Earth, its field gradient is weaker.

The solar tidal force is 46% as large as the lunar. More precisely, the lunar tidal acceleration (along the Moon–Earth axis, at the Earth's surface) is about $1.1 \times 10^{-7}\, g$, while the solar tidal acceleration (along the Sun–Earth axis, at the Earth's surface) is about $0.52 \times 10^{-7}\, g$, where g is the gravitational acceleration at the Earth's surface. Venus has the largest effect of the other planets, at 0.000113 times the solar effect.

The ocean's surface is closely approximated by an equipotential surface, (ignoring ocean currents) commonly referred to as the geoid. Since the gravitational force is equal to the potential's gradient, there are no tangential forces on such a surface, and the ocean surface is thus in gravitational equilibrium.

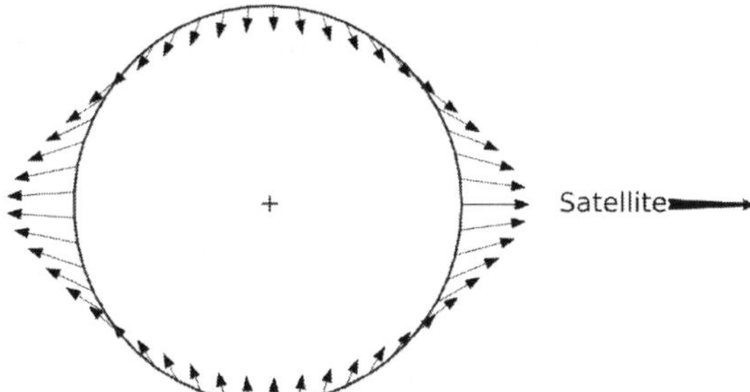

Figure: *The lunar gravity differential field at the Earth's surface is known as the tide-generating force. This is the primary mechanism that drives tidal action and explains two equipotential tidal bulges, accounting for two daily high waters.*

Now consider the effect of massive external bodies such as the Moon and Sun. These bodies have strong gravitational fields that diminish with distance in space and which act to alter the shape of an equipotential surface on the Earth. This deformation has a fixed spatial orientation relative to the influencing body. The Earth's rotation relative to this shape causes the daily tidal cycle. Gravitational forces follow an inverse-square law (force is inversely proportional to the square of the distance), but tidal forces are inversely proportional to the cube of the distance. The ocean surface moves because of the changing tidal equipotential, rising when the tidal potential is high, which occurs on the parts of the Earth nearest to and furthest from the Moon. When the tidal equipotential changes, the ocean surface is no longer aligned with it, so the apparent direction of the vertical shifts. The surface then experiences a down slope, in the direction that the equipotential has risen.

Laplace's Tidal Equations

Ocean depths are much smaller than their horizontal extent. Thus, the response to tidal forcing can be modelled using the Laplace tidal equations which incorporate the following features:

1. The vertical (or radial) velocity is negligible, and there is no vertical shear—this is a sheet flow.
2. The forcing is only horizontal (tangential).
3. The Coriolis effect appears as an inertial force (fictitious) acting laterally to the direction of flow and proportional to velocity.
4. The surface height's rate of change is proportional to the negative divergence of velocity multiplied by the depth. As the horizontal velocity stretches or compresses the ocean as a sheet, the volume thins or thickens, respectively.

The boundary conditions dictate no flow across the coastline and free slip at the bottom.

The Coriolis effect (inertial force) steers flows moving towards the equator to the west and flows moving away from the equator toward the east, allowing coastally trapped waves. Finally, a dissipation term can be added which is an analog to viscosity.

Amplitude and Cycle Time

The theoretical amplitude of oceanic tides caused by the moon is about 54 centimetres (21 in) at the highest point, which corresponds to the amplitude that would be reached if the ocean possessed a uniform depth, there were no landmasses, and the Earth were rotating in step with the moon's orbit. The sun similarly causes tides, of which the theoretical amplitude is about 25 centimetres (9.8 in) (46% of that of the moon) with a cycle time of 12 hours. At spring tide the two effects add to each other to a theoretical level of 79 centimetres (31 in), while at neap tide the theoretical level is reduced to 29 centimetres (11 in). Since the orbits of the Earth about the sun, and the moon about the Earth, are elliptical, tidal amplitudes change somewhat as a result of the varying Earth–sun and Earth–moon distances. This causes a variation in the tidal force and theoretical amplitude of about ±18% for the moon and ±5% for the sun. If both the sun and moon were at their closest positions and aligned at new moon, the theoretical amplitude would reach 93 centimetres (37 in).

Real amplitudes differ considerably, not only because of depth variations and continental obstacles, but also because wave propagation across the ocean has a natural period of the same order of magnitude as the rotation period: if there were no land masses, it would take about 30 hours for a long wavelength surface wave to propagate along the equator halfway around the Earth (by comparison, the Earth's lithosphere has a natural period of about 57 minutes). Earth tides, which raise and lower the bottom of the ocean, and the tide's own gravitational self attraction are both significant and further complicate the ocean's response to tidal forces.

Dissipation

Earth's tidal oscillations introduce dissipation at an average rate of about 3.75 terawatt. About 98% of this dissipation is by marine tidal movement. Dissipation arises as basin-scale tidal flows drive smaller-scale flows which experience turbulent dissipation. This tidal drag creates torque on the moon that gradually transfers angular momentum to its orbit, and a gradual increase in Earth–moon separation. The equal and opposite torque on the Earth correspondingly decreases its rotational velocity. Thus, over geologic time, the moon recedes from the Earth, at about 3.8 centimetres (1.5 in)/year, lengthening the terrestrial day. Day length has increased by about 2 hours in the last 600 million years. Assuming (as a crude approximation) that the deceleration rate has been constant, this would imply that 70 million years ago, day length was on the order of 1% shorter with about 4 more days per year.

Observation and Prediction

History

Figure: *Brouscon's Almanach of 1546: Compass bearings of high waters in the Bay of Biscay (left) and the coast from Brittany to Dover (right).*

Figure: *Brouscon's Almanach of 1546: Tidal diagrams "according to the age of the moon".*

From ancient times, tidal observation and discussion has increased in sophistication, first marking the daily recurrence, then tides' relationship to the sun and moon. Pytheas travelled to the British Isles about 325 BC and seems to be the first to have related spring tides to the phase of the moon.

In the 2nd century BC, the Babylonian astronomer, Seleucus of Seleucia, correctly described the phenomenon of tides in order to support his heliocentric theory. He correctly theorized that tides were caused by the moon, although he believed that the interaction was mediated by the pneuma. He noted that tides varied in time and strength in different parts of the world. According to Strabo (1.1.9), Seleucus was the first to link tides to the lunar attraction, and that the height of the tides depends on the moon's position relative to the sun.

The *Naturalis Historia* of Pliny the Elder collates many tidal observations, e.g., the spring tides are a few days after (or before) new and full moon and are highest around the equinoxes, though Pliny noted many relationships now regarded as fanciful. In his *Geography*, Strabo described tides in the Persian Gulf having their greatest range when the moon was furthest from the plane of the equator. All this despite the relatively small amplitude of Mediterranean basin tides. (The strong currents through the Euripus Strait and the Strait of Messina puzzled Aristotle.) Philostratus discussed tides in Book Five of *The Life of Apollonius of Tyana*. Philostratus mentions the moon, but attributes tides to "spirits". In Europe around 730 AD, the Venerable Bede described how the rising tide on one coast of the British Isles coincided with the fall on the other and described the time progression of high water along the Northumbrian coast.

The first tide table in China was recorded in 1056 AD primarily for visitors wishing to see the famous tidal bore in the Qiantang River. The first known British tide table is thought to be that of John Wallingford, who died Abbot of St. Albans in 1213, based on high water occurring 48 minutes later each day, and three hours earlier at the Thames mouth than upriver at London.

William Thomson (Lord Kelvin) led the first systematic harmonic analysis of tidal records starting in 1867. The main result was the building of a tide-predicting machine using a system of pulleys to add together six harmonic time functions. It was "programmed" by resetting gears and chains to adjust phasing and amplitudes. Similar machines were used until the 1960s.

The first known sea-level record of an entire spring–neap cycle was made in 1831 on the Navy Dock in the Thames Estuary. Many large ports had automatic tide gage stations by 1850.

William Whewell first mapped co-tidal lines ending with a nearly global chart in 1836. In order to make these maps consistent, he hypothesized the existence of amphidromes where co-tidal lines meet in the mid-ocean. These points of no tide were confirmed by measurement in 1840 by Captain Hewett, RN, from careful soundings in the North Sea.

Timing

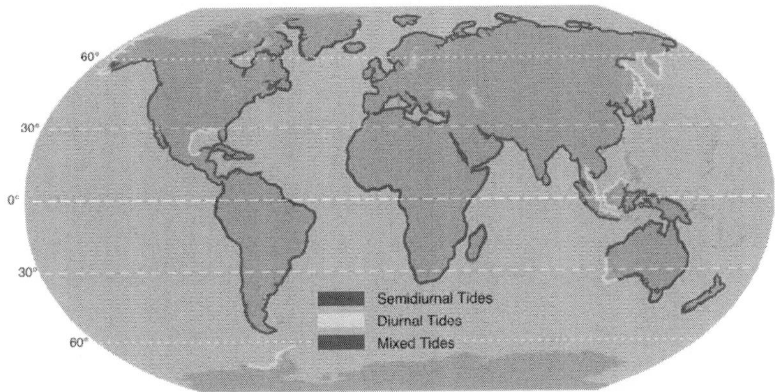

Figure: *The same tidal forcing has different results depending on many factors, including coast orientation, continental shelf margin, water body dimensions.*

The tidal forces due to the Moon and Sun generate very long waves which travel all around the ocean following the paths shown in co-tidal charts. The time when the crest of the wave reaches a port then gives the time of high water at the port. The time taken for the wave to travel around the ocean also means that there is a delay between the phases of the moon and their effect on the tide. Springs and neaps in the North Sea, for example, are two days behind the new/full moon and first/third quarter moon. This is called the tide's *age*.

The ocean bathymetry greatly influences the tide's exact time and height at a particular coastal point. There are some extreme cases; the Bay of Fundy, on the east coast of Canada, is often stated to have the world's highest tides because of its shape, bathymetry, and its distance from the continental shelf edge. Measurements made in November 1998 at Burntcoat Head in the Bay of Fundy recorded a maximum range of 16.3 metres (53 ft) and a highest predicted extreme of 17 metres (56 ft). Similar measurements made in March 2002 at Leaf Basin,

Ungava Bay in northern Quebec gave similar values (allowing for measurement errors), a maximum range of 16.2 metres (53 ft) and a highest predicted extreme of 16.8 metres (55 ft). Ungava Bay and the Bay of Fundy lie similar distances from the continental shelf edge, but Ungava Bay is free of pack ice for only about four months every year while the Bay of Fundy rarely freezes.

Southampton in the United Kingdom has a double high water caused by the interaction between the region's different tidal harmonics, caused primarily by the east/west orientation of the English Channel and the fact that when it is high water at Dover it is low water at Land's End (some 300 nautical miles distant) and vice versa. This is contrary to the popular belief that the flow of water around the Isle of Wight creates two high waters. The Isle of Wight is important, however, since it is responsible for the 'Young Flood Stand', which describes the pause of the incoming tide about three hours after low water.

Because the oscillation modes of the Mediterranean Sea and the Baltic Sea do not coincide with any significant astronomical forcing period, the largest tides are close to their narrow connections with the Atlantic Ocean. Extremely small tides also occur for the same reason in the Gulf of Mexico and Sea of Japan. Elsewhere, as along the southern coast of Australia, low tides can be due to the presence of a nearby amphidrome.

Analysis

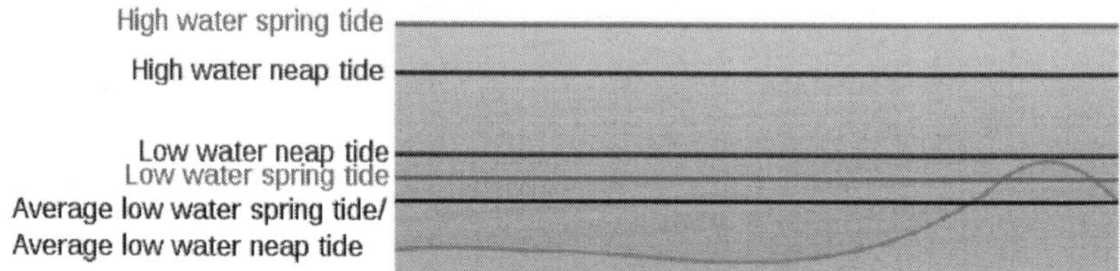

Figure: *A regular water level chart*

Isaac Newton's theory of gravitation first enabled an explanation of why there were generally two tides a day, not one, and offered hope for detailed understanding. Although it may seem that tides could be predicted via a sufficiently detailed knowledge of the instantaneous astronomical forcings, the actual tide at a given location is determined by astronomical forces accumulated over many days. Precise results require detailed knowledge of the shape of all the ocean basins—their bathymetry and coastline shape.

Current procedure for analysing tides follows the method of harmonic analysis introduced in the 1860s by William Thomson. It is based on the principle that the astronomical theories of the motions of sun and moon determine a large number of component frequencies, and at each frequency there is a component of force tending to produce tidal motion, but that at each place of interest on the Earth, the tides respond at each frequency with an amplitude and phase peculiar to that locality.

At each place of interest, the tide heights are therefore measured for a period of time sufficiently long (usually more than a year in the case of a new port not previously studied) to enable the response at each significant tide-generating frequency to be distinguished by analysis, and to extract the tidal constants for a sufficient number of the strongest known components of the astronomical tidal forces to enable practical tide prediction.

The tide heights are expected to follow the tidal force, with a constant amplitude and phase delay for each component. Because astronomical frequencies and phases can be calculated with certainty, the tide height at other times can then be predicted once the response to the harmonic components of the astronomical tide-generating forces has been found.

The main patterns in the tides are

- the twice-daily variation
- the difference between the first and second tide of a day
- the spring–neap cycle
- the annual variation

The *Highest Astronomical Tide* is the perigean spring tide when both the sun and the moon are closest to the Earth.

When confronted by a periodically varying function, the standard approach is to employ Fourier series, a form of analysis that uses sinusoidal functions as a *basis* set, having frequencies that are zero, one, two, three, etc. times the frequency of a particular fundamental cycle. These multiples are called *harmonics* of the fundamental frequency, and the process is termed harmonic analysis. If the basis set of sinusoidal functions suit the behaviour being modelled, relatively few harmonic terms need to be added. Orbital paths are very nearly circular, so sinusoidal variations are suitable for tides.

For the analysis of tide heights, the Fourier series approach has in practice to be made more elaborate than the use of a single frequency and its harmonics. The tidal patterns are decomposed into many sinusoids having many fundamental frequencies, corresponding (as in the lunar theory) to many different combinations of the motions of the Earth, the moon, and the angles that define the shape and location of their orbits.

For tides, then, *harmonic analysis* is not limited to harmonics of a single frequency. In other words, the harmonies are multiples of many fundamental frequencies, not just of the fundamental frequency of the simpler Fourier series approach. Their representation as a Fourier series having only one fundamental frequency and its (integer) multiples would require many terms, and would be severely limited in the time-range for which it would be valid.

The study of tide height by harmonic analysis was begun by Laplace, William Thomson (Lord Kelvin), and George Darwin. A.T. Doodson extended their work, introducing the *Doodson Number* notation to organise the hundreds of resulting terms. This approach has been the international standard ever since, and the complications arise as follows: the tide-raising force is notionally given by sums of several terms. Each term is of the form

$$A\cos(\omega t + p)$$

where A is the amplitude, ω is the angular frequency usually given in degrees per hour corresponding to t measured in hours, and p is the phase offset with regard to the astronomical state at time $t = 0$. There is one term for the moon and a second term for the sun. The phase p of the first harmonic for the moon term is called the lunitidal interval or high water interval. The next step is to accommodate the harmonic terms due to the elliptical shape of the orbits. Accordingly, the value of A is not a constant but also varying with time, slightly, about some average figure. Replace it then by $A(t)$ where A is another sinusoid, similar to the cycles and epicycles of Ptolemaic theory. Accordingly,

$$A(t) = A\big(1 + A_a \cos(\omega_a t + p_a)\big)$$

which is to say an average value A with a sinusoidal variation about it of magnitude A_a, with frequency ω_a and phase p_a. Thus the simple term is now the product of two cosine factors:

$$A\big[1 + A_a \cos(\omega_a t + p_a)\big]\cos(\omega t + p)$$

Given that for any x and y

$$\cos x \cos y = \frac{1}{2}\cos(x + y) + \frac{1}{2}\cos(x - y)$$

it is clear that a compound term involving the product of two cosine terms each with their own frequency is the same as *three* simple cosine terms that are to be added at the original frequency and also at frequencies which are the sum and difference of the two frequencies of the product term. (Three, not two terms, since the whole expression is $(1+\cos x)\cos y$.) Consider further that the tidal force on a location depends also on whether the moon (or the sun) is above or below the plane of the equator, and that these attributes have their own periods also incommensurable with a day and a month, and it is clear that many combinations result.

With a careful choice of the basic astronomical frequencies, the Doodson Number annotates the particular additions and differences to form the frequency of each simple cosine term.

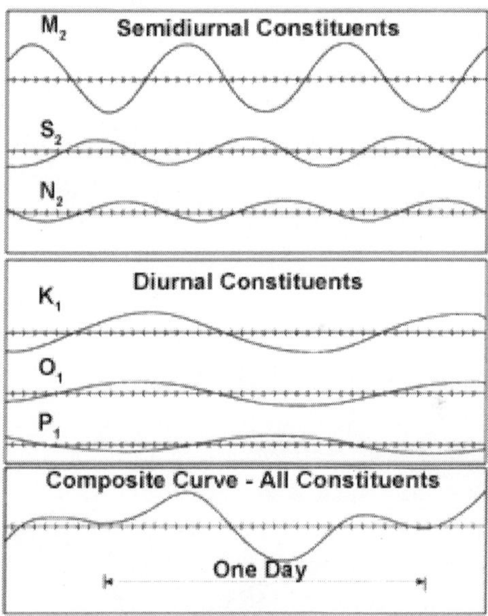

Figure: *Tidal prediction summing constituent parts.*

Remember that astronomical tides do *not* include weather effects. Also, changes to local conditions (sandbank movement, dredging harbour mouths, etc.) away from those prevailing at the measurement time affect the tide's actual timing and magnitude. Organisations quoting a "highest astronomical tide" for some location may exaggerate the figure as a safety factor against analytical uncertainties, distance from the nearest measurement point, changes since the last observation time, ground subsidence, etc., to avert liability should an engineering work be overtopped. Special care is needed when assessing the size of a "weather surge" by subtracting the astronomical tide from the observed tide.

Careful Fourier data analysis over a nineteen-year period (the *National Tidal Datum Epoch* in the U.S.) uses frequencies called the *tidal harmonic constituents.* Nineteen years is preferred because the Earth, moon and sun's relative positions repeat almost exactly in the Metonic cycle of 19 years, which is long enough to include the 18.613 year lunar nodal tidal constituent. This analysis can be done using only the knowledge of the forcing *period,* but without detailed understanding of the mathematical derivation, which means that useful tidal tables have been constructed for centuries.

The resulting amplitudes and phases can then be used to predict the expected tides. These are usually dominated by the constituents near 12 hours (the *semi-diurnal* constituents), but there are major constituents near 24 hours (*diurnal*) as well. Longer term constituents are 14 day or *fortnightly*, monthly, and semiannual. Semi-diurnal tides dominated coastline, but some areas such as the South China Sea and the Gulf of Mexico are primarily diurnal. In the semi-diurnal areas, the primary constituents M_2 (lunar) and S_2 (solar) periods differ slightly, so that the relative phases, and thus the amplitude of the combined tide, change fortnightly (14 day period).

In the M_2 plot above, each cotidal line differs by one hour from its neighbours, and the thicker lines show tides in phase with equilibrium at Greenwich. The lines rotate around the amphidromic points counterclockwise in the northern hemisphere so that from Baja California Peninsula to Alaska and from France to Ireland the M_2 tide propagates northward. In the southern hemisphere this direction is clockwise. On the other hand M_2 tide propagates counterclockwise around New Zealand, but this is because the islands act as a dam and permit the tides to have different heights on the islands' opposite sides. (The tides do propagate northward on the east side and southward on the west coast, as predicted by theory.)

The exception is at Cook Strait where the tidal currents periodically link high to low water. This is because cotidal lines 180° around the amphidromes are in opposite phase, for example high water across from low water at each end of Cook Strait. Each tidal constituent has a different pattern of amplitudes, phases, and amphidromic points, so the M_2 patterns cannot be used for other tide components.

Example Calculation

Because the moon is moving in its orbit around the earth and in the same sense as the Earth's rotation, a point on the earth must rotate slightly further to catch up so that the time between semidiurnal tides is not twelve but 12.4206 hours—a bit over twenty-five minutes extra. The two peaks are not equal. The two high tides a day alternate in maximum heights: lower high (just under three feet), higher high (just over three feet), and again lower high. Likewise for the low tides.

When the Earth, moon, and sun are in line (sun–Earth–moon, or sun–moon–Earth) the two main influences combine to produce spring tides; when the two forces are opposing each other as when the angle moon–Earth–sun is close to ninety degrees, neap tides result. As the moon moves around its orbit it changes from north of the equator to south of the equator.

The alternation in high tide heights becomes smaller, until they are the same (at the lunar equinox, the moon is above the equator), then redevelop but with the other polarity, waxing to a maximum difference and then waning again.

Current

The tides' influence on current flow is much more difficult to analyse, and data is much more difficult to collect. A tidal height is a simple number which applies to a wide region simultaneously. A flow has both a magnitude and a direction, both of which can vary substantially with depth and over short distances due to local bathymetry. Also, although a water channel's centre is the most useful measuring site, mariners object when current-measuring equipment obstructs waterways. A flow proceeding up a curved channel is the same flow, even though its direction varies continuously along the channel. Surprisingly, flood and ebb flows are often not in opposite directions. Flow direction is determined by the upstream channel's shape, not the downstream channel's shape. Likewise, eddies may form in only one flow direction.

Nevertheless, current analysis is similar to tidal analysis: in the simple case, at a given location the flood flow is in mostly one direction, and the ebb flow in another direction. Flood velocities are given positive sign, and ebb velocities negative sign. Analysis proceeds as though these are tide heights.

In more complex situations, the main ebb and flood flows do not dominate. Instead, the flow direction and magnitude trace an ellipse over a tidal cycle (on a polar plot) instead of along the ebb and flood lines. In this case, analysis might proceed along pairs of directions, with the primary and secondary directions at right angles. An alternative is to treat the tidal flows as complex numbers, as each value has both a magnitude and a direction.

Tide flow information is most commonly seen on nautical charts, presented as a table of flow speeds and bearings at hourly intervals, with separate tables for spring and neap tides. The timing is relative to high water at some harbour where the tidal behaviour is similar in pattern, though it may be far away.

As with tide height predictions, tide flow predictions based only on astronomical factors do not incorporate weather conditions, which can *completely* change the outcome.

The tidal flow through Cook Strait between the two main islands of New Zealand is particularly interesting, as the tides on each side of the strait are almost exactly out of phase, so that one side's high water is simultaneous with the other's low water. Strong currents result, with almost zero tidal height change in the strait's centre.

Yet, although the tidal surge normally flows in one direction for six hours and in the reverse direction for six hours, a particular surge might last eight or ten hours with the reverse surge enfeebled. In especially boisterous weather conditions, the reverse surge might be entirely overcome so that the flow continues in the same direction through three or more surge periods.

A further complication for Cook Strait's flow pattern is that the tide at the north side (e.g. at Nelson) follows the common bi-weekly spring–neap tide cycle (as found along the west side of the country), but the south side's tidal pattern has only *one* cycle per month, as on the east side: Wellington, and Napier.

The graph of Cook Strait's tides shows separately the high water and low water height and time, through November 2007; these are *not* measured values but instead are calculated from tidal parameters derived from years-old measurements. Cook Strait's nautical chart offers tidal current information. For instance the January 1979 edition for 41°13·9'S 174°29·6'E (north west of Cape Terawhiti) refers timings to Westport while the January 2004 issue refers to Wellington.

Near Cape Terawhiti in the middle of Cook Strait the tidal height variation is almost nil while the tidal current reaches its maximum, especially near the notorious Karori Rip. Aside from weather effects, the actual currents through Cook Strait are influenced by the tidal height differences between the two ends of the strait and as can be seen, only one of the two spring tides at the north end (Nelson) has a counterpart spring tide at the south end (Wellington), so the resulting behaviour follows neither reference harbour.

Power Generation

Tidal energy can be extracted by two means: inserting a water turbine into a tidal current, or building ponds that release/admit water through a turbine. In the first case, the energy amount is entirely determined by the timing and tidal current magnitude. However, the best currents may be unavailable because the turbines would obstruct ships. In the second, the impoundment dams are expensive to construct, natural water cycles are completely disrupted, ship navigation is disrupted. However, with multiple ponds, power can be generated at chosen times. So far, there are few installed systems for tidal power generation (most famously, La Rance at Saint Malo, France) which face many difficulties. Aside from environmental issues, simply withstanding corrosion and biological fouling pose engineering challenges.

Tidal power proponents point out that, unlike wind power systems, generation levels can be reliably predicted, save for weather effects. While some generation is possible for most of the tidal cycle, in practice turbines lose efficiency at lower operating rates. Since the power available from a flow is proportional to the cube of the flow speed, the times during which high power generation is possible are brief.

Navigation

DATUMS

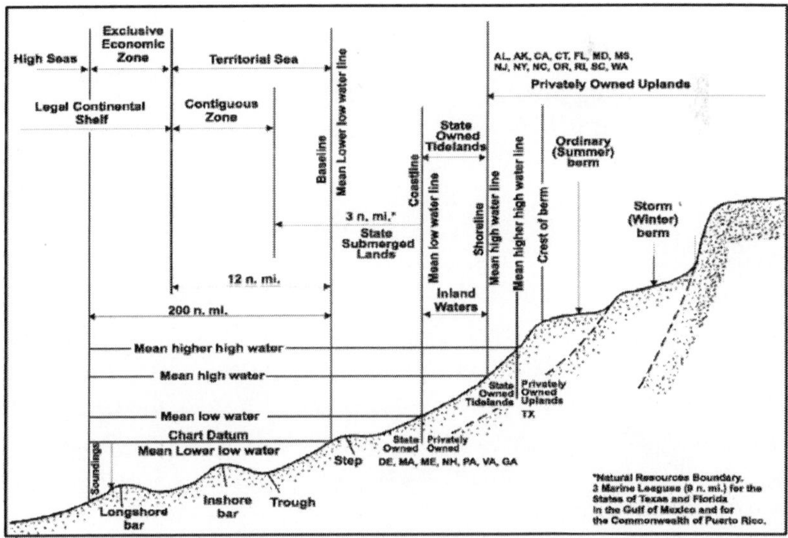

Figure: Civil and maritime uses of tidal data

Tidal flows are important for navigation, and significant errors in position occur if they are not accommodated. Tidal heights are also important; for example many rivers and harbours have a shallow "bar" at the entrance which prevents boats with significant draft from entering at low tide. Until the advent of automated navigation, competence in calculating tidal effects was important to naval officers. The certificate of examination for lieutenants in the Royal Navy once declared that the prospective officer was able to "shift his tides".

Tidal flow timings and velocities appear in *tide charts* or a tidal stream atlas. Tide charts come in sets. Each chart covers a single hour between one high water and another (they ignore the leftover 24 minutes) and show the average tidal flow for that hour. An arrow on the tidal chart indicates the direction and the average flow speed (usually in knots) for spring and neap tides. If a tide chart is not available, most nautical charts have "tidal diamonds" which relate specific points on the chart to a table giving tidal flow direction and speed.

The standard procedure to counteract tidal effects on navigation is to

(1) calculate a "dead reckoning" position (or DR) from travel distance and direction,

(2) mark the chart (with a vertical cross like a plus sign) and

(3) draw a line from the DR in the tide's direction.

The distance the tide moves the boat along this line is computed by the tidal speed, and this gives an "estimated position" or EP (traditionally marked with a dot in a triangle).

Nautical charts display the water's "charted depth" at specific locations with "soundings" and the use of bathymetric contour lines to depict the submerged surface's shape. These depths are relative to a "chart datum", which is typically the water level at the lowest possible astronomical tide (although other datums are commonly used, especially historically, and tides may be lower or higher for meteorological reasons) and are therefore the minimum possible water depth during the tidal cycle. "Drying heights" may also be shown on the chart, which are the heights of the exposed seabed at the lowest astronomical tide. Tide tables list each day's high and low water heights and times. To calculate the actual water depth, add the charted depth to the published tide height. Depth for other times can be derived from tidal curves published for major ports. The rule of twelfths can suffice if an accurate curve is not available. This approximation presumes that the increase in depth in the six hours between low and high water is: first hour — 1/12, second — 2/12, third — 3/12, fourth — 3/12, fifth — 2/12, sixth — 1/12.

Biological Aspects

Intertidal Ecology

Intertidal ecology is the study of intertidal ecosystems, where organisms live between the low and high water lines. At low water, the intertidal is exposed (or 'emersed') whereas at high water, the intertidal is underwater (or 'immersed'). Intertidal ecologists therefore study the interactions between intertidal organisms and their environment, as well as among the different species. The most important interactions may vary according to the type of intertidal community. The broadest classifications are based on substrates — rocky shore or soft bottom. Intertidal organisms experience a highly variable and often hostile environment, and have adapted to cope with and even exploit these conditions. One easily visible feature is vertical zonation, in which the community divides into distinct horizontal bands of specific species at each elevation above low water. A species' ability to cope with desiccation determines its upper limit, while competition with other species sets its lower limit. Humans use intertidal regions for food and recreation. Overexploitation can damage intertidals directly. Other anthropogenic actions such as introducing invasive species and climate change have large negative effects. Marine Protected Areas are one option communities can apply to protect these areas and aid scientific research.

Biological Rhythms

The approximately fortnightly tidal cycle has large effects on intertidal and marine organisms. Hence their biological rhythms tend to occur in rough multiples of this period. Many other animals such as the vertebrates, display similar rhythms. Examples include gestation and egg hatching. In humans, the menstrual cycle lasts roughly a lunar month, an even multiple of the tidal period. Such parallels at least hint at the common descent of all animals from a marine ancestor.

Other Tides

When oscillating tidal currents in the stratified ocean flow over uneven bottom topography, they generate internal waves with tidal frequencies. Such waves are called *internal tides*. Shallow areas in otherwise open water can experience rotary tidal currents, flowing in directions that continually change and thus the flow direction (not the flow) completes a full rotation in 12½ hours (for example, the Nantucket Shoals).

In addition to oceanic tides, large lakes can experience small tides and even planets can experience *atmospheric tides* and *Earth tides*.

These are continuum mechanical phenomena. The first two take place in fluids. The third affects the Earth's thin solid crust surrounding its semi-liquid interior (with various modifications).

Lake Tides

Large lakes such as Superior and Erie can experience tides of 1 to 4 cm, but these can be masked by meteorologically induced phenomena such as seiche. The tide in Lake Michigan is described as 0.5 to 1.5 inches (13 to 38 mm) or 1¾ inches.

Atmospheric Tides

Atmospheric tides are negligible at ground level and aviation altitudes, masked by weather's much more important effects. Atmospheric tides are both gravitational and thermal in origin and are the dominant dynamics from about 80 to 120 kilometres (50 to 75 mi), above which the molecular density becomes too low to support fluid behaviour.

Earth Tides

Earth tides or terrestrial tides affect the entire Earth's mass, which acts similarly to a liquid gyroscope with a very thin crust. The Earth's crust shifts (in/out, east/west, north/south) in response to lunar and solar gravitation, ocean tides, and atmospheric loading. While negligible for most human activities, terrestrial tides' semi-diurnal amplitude can reach about 55 centimetres (22 in) at the equator—15 centimetres (5.9 in) due to the sun—which is important in GPS calibration and VLBI measurements. Precise astronomical angular measurements require knowledge of the Earth's rotation rate and nutation, both of which are influenced by Earth tides. The semi-diurnal M_2 Earth tides are nearly in phase with the moon with a lag of about two hours.

Some particle physics experiments must adjust for terrestrial tides. For instance, at CERN and SLAC, the very large particle accelerators account for terrestrial tides. Among the relevant effects are circumference deformation for circular accelerators and particle beam energy. Since tidal forces generate currents in conducting fluids in the Earth's interior, they in turn affect the Earth's magnetic field. Earth tides have also been linked to the triggering of earthquakes.

Chapter 9

Weightlessness

Weightlessness, or an absence of 'weight', is in fact an absence of stress and strain resulting from externally applied mechanical contact-forces, typically normal forces from floors, seats, beds, scales, and the like. Counterintuitively, a uniform gravitational field does not by itself cause stress or strain, and a body in free fall in such an environment experiences no g-force acceleration and feels weightless. This is also termed "zero-g" where the term is most correctly understood as meaning "zero g-force."

When bodies are acted upon by non-gravitational forces, as in a centrifuge, a rotating space station, or within a space ship with rockets firing, a sensation of weight is produced, as the contact forces from the moving structure act to overcome the body's inertia. In such cases, a sensation of weight, in the sense of a state of stress can occur, even if the gravitational field was zero. In such cases, g-forces are felt, and bodies are not weightless.

When the gravitational field is non-uniform, a body in free fall suffers tidal effects and is not stress-free. Near a black hole, such tidal effects can be very strong. In the case of the Earth, the effects are minor, especially on objects of relatively small dimension (such as the human body or a spacecraft) and the overall sensation of weightlessness in these cases is preserved. This condition is known as microgravity and it prevails in orbiting spacecraft.

Weightlessness in Newtonian Mechanics

In Newtonian mechanics the term "weight" is given two distinct interpretations by engineers.

Weight$_1$: Under this interpretation, the "weight" of a body is the gravitational force exerted on the body and this is the notion of weight that prevails in engineering. Near the surface of the earth, a body whose mass is 1 kg has a weight of approximately 10 N, independent of its state of motion, *free fall, or not*. Weightlessness in this sense can be achieved by removing the body far away from the source of gravity. It can also be attained by placing the body at a neutral point between two gravitating masses.

Weight$_2$: Weight can also be interpreted as that quantity which is measured when one uses scales. What is being measured there is the force exerted *by* the body on the scales. In a standard weighing operation, the body being weighed is in a state of equilibrium as a result of a force exerted on it by the weighing machine cancelling the gravitational field. By Newton's 3rd law, there is an equal and opposite force exerted *by* the body on the machine. *This* force is called weight$_2$. The force is *not* gravitational. Typically, it is a contact force and not uniform across the mass of the body. If the body is placed on the scales in a lift (an elevator) in free fall in pure uniform gravity, the scale would read zero, and the body said to be weightless i.e. its weight$_2$ = 0. This describes the condition in which the body is stress free and undeformed. This is the weightlessness in *free fall in a uniform gravitational field*. (The situation is more complicated when the gravitational field is not uniform, or, when a body is subject to multiple forces which may, for instance, cancel each other and produce a state of stress albeit weight$_2$ being zero.)

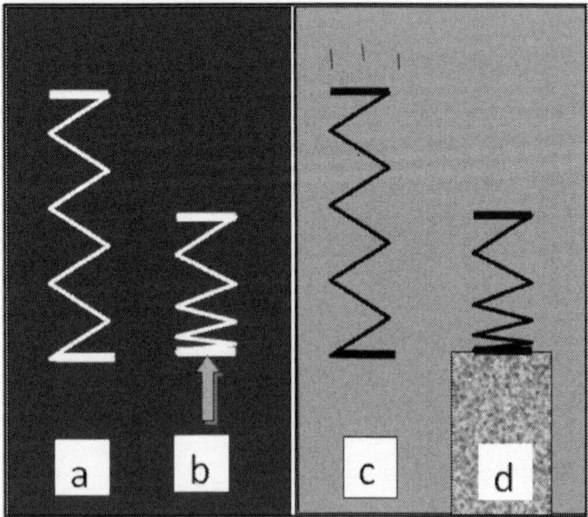

Figure: *In the left half, the spring is far away from any gravity source. In the right half, it is in a uniform gravitation field. a) Zero gravity and weightless b) Zero gravity but not weightless (Spring is rocket propelled) c) Spring is in free fall and weightless d) Spring rests on a plinth and has both weight$_1$ and weight$_2$.*

To sum up, we have two notions of weight of which weight$_1$ is dominant. Yet 'weightlessness' is typically exemplified not by absence of weight$_1$ but by the absence of stress associated with weight$_2$. This is the intended sense of weightlessness in what follows below.

A body is stress free, exerts zero weight$_2$, when the only force acting on it is weight$_1$ as when in free fall in a uniform gravitational field. Without subscripts, one ends up with the odd-sounding conclusion that a body is weightless when the only force acting on it is its weight.

The apocryphal apple that fell on Newton's head can be used to illustrate the issues involved. An apple weighs approximately 1 Newton. This is the weight$_1$ of the apple and is considered to be a constant even while it is falling. During that fall, its weight$_2$ however is zero: ignoring air resistance, the apple is stress free. When it hits Newton, the sensation felt by Newton would depend upon the height from which the apple falls and weight$_2$ of the apple at the moment of impact may be many times greater than 1 N. It was great enough—in the story—to make the great man invent the theory of gravity. It is this weight$_2$ which distorts the apple. On its way down, the apple in its free fall does not suffer any distortion as the gravitational field is uniform.

Stress During Free Fall

1. In a uniform gravitational field: Consider any cross-section dividing the body into two parts. Both parts have the same acceleration and the force exerted on each is supplied by the external source of the field. There is no force exerted by one part on the other. Stress at the cross-section is zero. Weight$_2$ is zero.
2. In a non-uniform gravitational field: Under gravity alone, one part of the body may have a different acceleration from another part. This would tend to deform the body and generate internal stresses if the body resists deformation. Weight$_2$ is not 0.

Throughout this discussion on using stress as an indicator of weight, any *pre-stress* which may exist within a body caused by a force exerted on one part by another is not relevant. The only relevant stresses are those generated by *external* forces applied to the body.

The definition and use of 'weightlessness' is difficult unless it is understood that the sensation of "weight" in everyday terrestrial experience results not from gravitation acting alone (which is not felt), but instead by the mechanical forces that resist gravity. An object in a straight free fall, or in a more complex inertial trajectory of free fall (such as within a reduced gravity aircraft or inside a space station), all experience weightlessness, since they do not experience the mechanical forces that cause the sensation of weight.

Force Fields Other than Gravity

As noted above, weightlessness occurs when 1. no force acts on the object 2. uniform gravity acts solely by itself. For the sake of completeness, a 3rd minor possibility has to be added. This is that a body may be subject to a field which is not gravitational but such that the force on the object is *uniformly distributed* across the object's mass. An electrically charged body, uniformly charged, in a uniform electric field is a possible example. Electric charge here replaces the usual gravitational charge. Such a body would then be stress free and be classed as weightless. Various types of levitation may fall into this category, at least approximately.

Weightlessness and Proper Acceleration

A body in free fall (which by definition entails no aerodynamic forces) near the surface of the earth has an acceleration approximately equal to 10 m s^{-2} with respect to a coordinate frame tied to the earth. If the body is in a freely falling lift and subject to no pushes or pulls from the lift or its contents, the acceleration with respect to the lift would be zero. If on the other hand, the body is subject to forces exerted by other bodies within the lift, it will have an acceleration with respect to the freely falling lift. This acceleration which is not due to gravity is called "proper acceleration". On this approach, weightlessness holds when proper acceleration is zero.

How to Avoid Weightlessness

Weightlessness is in contrast with current human experiences in which a non-uniform force is acting, such as:

- standing on the ground, sitting in a chair on the ground, etc., where gravity is countered by the support force of the ground,
- flying in a plane, where a support force is transmitted from the lift the wings provide,
- during atmospheric reentry, or during the use of a parachute, when atmospheric drag decelerates a vehicle,
- during an orbital maneuver in a spacecraft, or during the launch phase, when rocket engines provide thrust.

In cases where an object is not weightless, as in the above examples, a force acts non-uniformly on the object in question. Aero-dynamic lift, drag, and thrust are all non-uniform forces (they are applied at a point or surface, rather than acting on the entire mass of an object), and thus create the phenomenon of weight. This non-uniform force may also be transmitted to an object at the point of contact with a second object, such as the contact between the surface of the Earth and one's feet, or between a parachute harness and one's body.

Tidal Forces

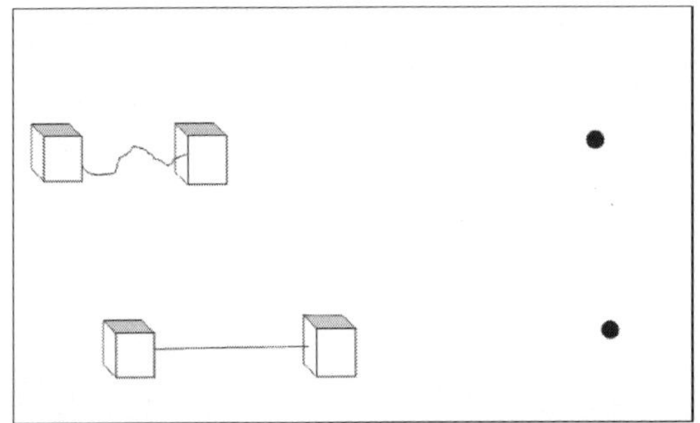

Figure: *Two rigid cubes joined by an elastic string in free fall near a black hole. The string stretches as the body falls to the right.*

Tidal forces arise when the gravitational field is not uniform and *gravitation gradients* exist. Such indeed is the norm and strictly speaking any object of finite size even in free-fall is subject to tidal effects. These are impossible to remove by inertial motion, except at one single nominated point of the body. The Earth is in free fall but the presence of tides indicates that it is in a non-uniform gravitational field. This non-uniformity is more due to the moon than the sun. The total gravitational field due to the sun is much stronger than that of the moon but it has a minor tidal effect compared with that of the moon because of the relative distances involved. $Weight_1$ of the earth is essentially due to the sun's gravity. But its state of stress and deformation, represented by the tides, is more due to non uniformity in the gravitational field of the nearby moon. When the size of a region being considered is small relative to its distance from the gravitating mass the assumption of uniform gravitational field holds to a good approximation. Thus a person is small relative to the radius of Earth and the field for a person at the surface of the earth is approximately uniform. The field is strictly not uniform and is responsible for the phenomenon of microgravity. Objects near a black hole are subject to a highly non-uniform gravitational field.

Frames of Reference

In all inertial reference frames, while weightlessness is experienced, Newton's first law of motion is obeyed locally *within the frame*. Inside the frame (for example, inside an orbiting ship or free-falling elevator), unforced objects keep their velocity relative to the frame. Objects not in contact with other objects "float" freely. If the inertial trajectory is influenced by gravity, the reference frame will be an accelerated frame as seen from a position outside the gravitational attraction, and (seen from far away) the objects in the frame (elevator, etc.) will appear to be under the influence of a force (the so-called force of gravity). As noted, objects subject solely to gravity do not feel its effects. Weightlessness can thus be realised for short periods of time in an airplane following a specific elliptic flight path, often mistakenly called a parabolic flight. It is simulated poorly, with many differences, in neutral buoyancy conditions, such as immersion in a tank of water.

Zero-g, "Zero Gravity", Accelerometres

Zero-g is an alternative term for weightlessness and holds for instance in a freely falling lift. Zero-g is subtly different from the complete absence of gravity, something which is impossible due to the presence of gravity everywhere in the universe. "Zero-gravity" may also be used to mean effective weightlessness, neglecting tidal effects. *Microgravity* (or μg) is used to refer to situations that are substantially weightless but where g-force stresses within objects due to tidal effects, as discussed above, are around a millionth of that at the Earth's surface. Accelerometres can only detect g-force i.e. weight_2 (= mass × proper acceleration). They cannot detect the acceleration associated with free fall.

Sensation of Weight

Humans experience their own body weight as a result of this supporting force, which results in a normal force applied to a person by the surface of a supporting object, on which the person is standing or sitting. In the absence of this force, a person would be in free-fall, and would experience weightlessness. It is the transmission of this reaction force through the human body, and the resultant compression and tension of the body's tissues, that results in the sensation of weight.

Because of the distribution of mass throughout a person's body, the magnitude of the reaction force varies between a person's feet and head. At any horizontal cross-section of a person's body (as with any column), the size of the compressive force being resisted by the tissues below the cross-section is equal to the weight of the portion of the body above the cross-section. In the pose adopted in the accompanying illustration, the shoulders carry the weight of the outstretched arms and are subject to a considerable torque.

A Common Misconception

A common conception about spacecraft orbiting the earth is that they are operating in a gravity free environment. Although there is a way of making sense of this within the physics of Einstein's general relativity, within Newtonian physics, this is technically inaccurate.

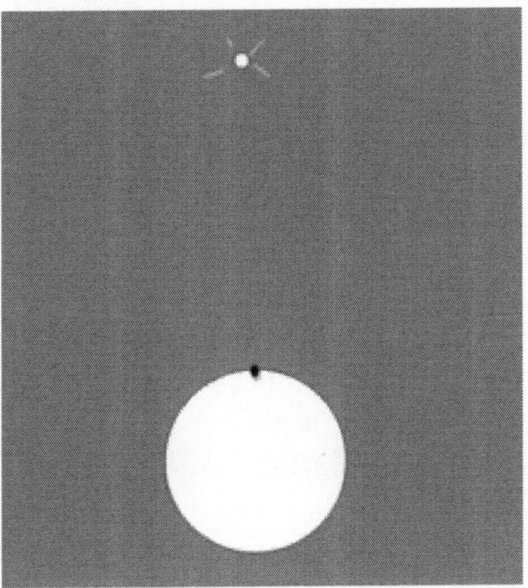

Figure: *A geostationary satellite above a marked spot on the Equator. An observer on the marked spot will see the satellite remain directly overhead unlike the other heavenly objects which sweep across the sky.*

Spacecraft are held in orbit by the gravity of the planet which they are orbiting. In Newtonian physics, the sensation of weightlessness experienced by astronauts is not the result of there being zero gravitational acceleration (as seen from the Earth), but of there being no g-force that an astronaut can feel because of the free-fall condition, and also there being zero difference between the acceleration of the spacecraft and the acceleration of the astronaut. Space journalist James Oberg explains the phenomenon this way:

The myth that satellites remain in orbit because they have "escaped Earth's gravity" is perpetuated further (and falsely) by almost universal misuse of the word "zero gravity" to describe the free-falling conditions aboard orbiting space vehicles. Of course, this isn't true; gravity still exists in space. It keeps satellites from flying straight off into interstellar emptiness. What's missing is "weight", the resistance of gravitational attraction by an anchored structure or a counterforce. Satellites stay in space because of their tremendous horizontal speed, which allows them — while being unavoidably pulled toward Earth by gravity — to fall "over the horizon." The ground's curved withdrawal along the Earth's round surface offsets the satellites' fall toward the ground. Speed, not position or lack of gravity, keeps satellites in orbit around the earth.

A geostationary satellite is of special interest in this context. Unlike other objects in the sky which rise and set, an object in a geostationary orbit appears motionless in the sky, apparently defying gravity. In fact, it is in a circular equatorial orbit with a period of one day.

Relativity

To a modern physicist working with Einstein's general theory of relativity, the situation is even more complicated than is suggested above. Einstein's theory suggests that it actually is valid to consider that objects in inertial motion (such as falling in an elevator, or in a parabola in an airplane, or orbiting a planet) can indeed be considered to experience a local loss of the gravitational field in their rest frame. Thus, in the point of view (or frame) of the astronaut or orbiting ship, there actually is nearly-zero proper acceleration (the acceleration felt locally), just as would be the case far out in space, away from any mass. It is thus valid to consider that most of the gravitational field in such situations is actually absent from the point of view of the falling observer, just as the colloquial view suggests.

However, this loss of gravity for the falling or orbiting observer, in Einstein's theory, is due to the falling motion itself, and (again as in Newton's theory) not due to increased distance from the Earth. However, the gravity nevertheless is considered to be absent. In fact, Einstein's realisation that a pure gravitational interaction cannot be felt, if all other forces are removed, was the key insight to leading him to the view that the gravitational "force" can in some ways be viewed as non-existent. Rather, objects tend to follow geodesic paths in curved space-time, and this is "explained" as a force, by "Newtonian" observers who assume that space-time is "flat," and thus do not have a reason for curved paths (i.e., the "falling motion" of an object near a gravitational source).

In the theory of general relativity, the only gravity which remains for the observer following a falling path or "inertial" path near a gravitating body, is that which is due to non-uniformities which remain in the gravitational field, even for the falling observer. This non-uniformity, which is a simple tidal effect in Newtonian dynamics, constitutes the "microgravity" which is felt by all spacially-extended objects falling in any natural gravitational field that originates from a compact mass. The reason for these tidal effects is that such a field will have its origin in a centralized place (the compact mass), and thus will diverge, and vary slightly in strength, according to distance from the mass. It will thus vary across the width of the falling or orbiting object. Thus, the term "microgravity," an overly technical term from the Newtonian view, is a valid and descriptive term in the general relativistic (Einsteinian) view.

Microgravity

The term micro-g environment (also μg, often referred to by the term microgravity) is more or less a synonym of weightlessness and *zero-G*, but indicates that g-forces are not quite zero, just very small.

A "stationary" micro-g environment would require travelling far enough into deep space so as to reduce the effect of gravity by attenuation to almost zero. This is the simplest in conception, but requires travelling an enormous distance, rendering it most impractical. For example, to reduce the gravity of the Earth by a factor of one million, one needs to be at a distance of 6 million km from the Earth, but to reduce the gravity of the Sun to this amount one has to be at a distance of 3.7 billion km. (The gravity due to the rest of the Milky Way is already smaller than one millionth of the gravity on Earth, so we do not need to move away further from its centre). Thus it is not impossible, but it has only been achieved so far by four interstellar probes (Voyager 1 and 2, part of the Voyager program, Pioneer 10 and 11 part of the Pioneer program) and they did not return to Earth. To reduce the gravity to one thousandth of that on Earth one needs to be at a distance of 200,000 km.

Location	*Gravity due to*			*Total*
	Earth	*Sun*	*rest of Milky Way*	
Earth's surface	9.81 m/s²	6 mm/s²	200 pm/s² = 6 mm/s/yr	9.81 m/s²
Low Earth orbit	9 m/s²	6 mm/s²	200 pm/s²	9 m/s²
200,000 km from Earth	10 mm/s²	6 mm/s²	200 pm/s²	up to 12 mm/s²
6 million km from Earth	10 µm/s²	6 mm/s²	200 pm/s²	6 mm/s²
3.7 billion km from Earth	29 pm/s²	10 µm/s²	200 pm/s²	10 µm/s²
Voyager 1 (17 billion km from Earth)	1 pm/s²	500 nm/s²	200 pm/s²	500 nm/s²
0.1 light-year from Earth	400 am/s²	200 pm/s²	200 pm/s²	up to 400 pm/s²

From stationarity the gravity from "the rest of the Milky Way" would cause a free fall, covering a distance of 100 pm in one second, 360 nm in one minute, 1.3 mm in one hour, 70 cm in one day, 37 m in one week, 100 km in one year, and 10,000 km in 10 years (at a speed at that last location of 6 cm/s).

At a distance relatively close to Earth (less than 3000 km), gravity is only slightly reduced. As an object orbits a body such as the Earth, gravity is still attracting objects towards the Earth and the object is accelerated downward at almost 1g. Because the objects are typically moving laterally with respect to the surface at such immense speeds, the object will not lose altitude because of the curveture of the Earth. When viewed from an orbiting observer, other close objects in space appear to be floating because everything is being pulled towards Earth at the same speed, but also moving forward as the Earth's surface 'falls' away below. All these objects are in free fall, not zero gravity.

Weightless and Reduced Weight Environments

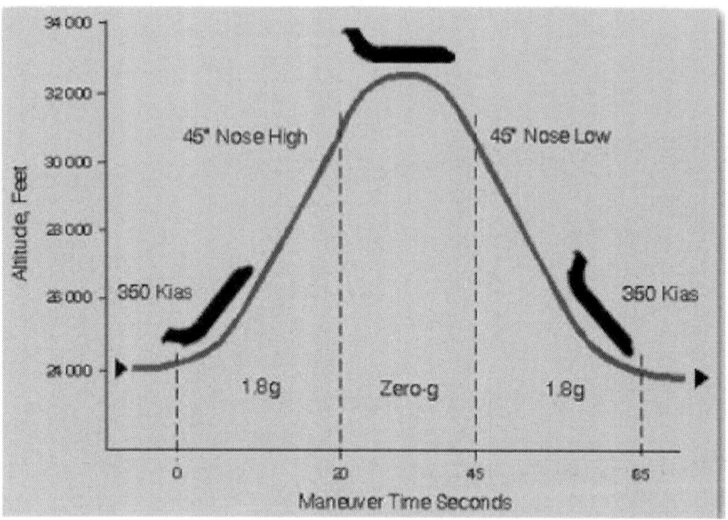

Figure: *Zero gravity flight maneuver*

Reduced Weight in Aircraft

Airplanes have been used since 1959 to provide a nearly weightless environment in which to train astronauts, conduct research, and film motion pictures. Such aircraft are commonly referred by the nickname "Vomit Comet".

To create a weightless environment, the airplane flies in a six-mile long parabolic arc, first climbing, then entering a powered dive. During the arc, the propulsion and steering of the aircraft are controlled such that the drag (air resistance) on the plane is cancelled out, leaving the plane to behave as it would if it were free-falling in a vacuum. During this period, the plane's occupants experience about 25 seconds of weightlessness, before experiencing about 25 seconds of 2 *g* acceleration (twice their normal weight) during the pull-out from the parabola. A typical flight lasts around two hours, during which 50 parabolae are flown.

NASA's Reduced Gravity Aircraft

Versions of such airplanes have been operated by NASA's Reduced Gravity Research Program since 1973, where the unofficial nickname originated. NASA later adopted the official nickname 'Weightless Wonder' for publication. NASA's current Reduced Gravity Aircraft, "Weightless Wonder VI", a McDonnell Douglas C-9, is based at Ellington Field (KEFD), near Lyndon B. Johnson Space Centre.

NASA's Microgravity University - Reduced Gravity Flight Opportunities Plan, also known as the Reduced Gravity Student Flight Opportunities Program, allows teams of undergraduates to submit a microgravity experiment proposal. If selected, the teams design and implement their experiment, and students are invited to fly on NASA's Vomit Comet.

Figure: *NASA's KC-135A plane ascending for a zero gravity maneuver*

European Space Agency A300 Zero-G

The European Space Agency flies parabolic flights on a specially-modified Airbus A300 B2 aircraft, in order to perform research in microgravity. ESA flies *campaigns* of three flights on consecutive days, each flying about 30 parabolas, for a total of about 10 minutes of weightlessness per flight. The ESA campaigns are currently operated from Bordeaux - Mérignac Airport in France by the company Novespace, while the aircraft is operated by DGA Essais en Vol. The first ESA Zero-G flights were in 1984, using a NASA KC-135 aircraft in Houston, Texas. As of May 2010, the ESA has flown 52 campaigns and also 9 student parabolic flight campaigns. Other aircraft it has used include the Russian Ilyushin Il-76 MDK and French Caravelle.

Others

The Zero Gravity Corporation, founded in 1993 by Peter Diamandis, Byron Lichtenberg, and Ray Cronise, operates a modified Boeing 727 which flies parabolic arcs to create 25–30 seconds of weightlessness. Flights may be purchased for both tourism and research purposes.

Ground-based Drop Facilities

Ground-based facilities that produce weightless conditions for research purposes are typically referred to as drop tubes or drop towers.

NASA's Zero Gravity Research Facility, located at the Glenn Research Centre in Cleveland, Ohio, is a 145-metre vertical shaft, largely below the ground, with an integral vacuum drop chamber, in which an experiment vehicle can have a free fall for a duration of 5.18 seconds, falling a distance of 132 metres. The experiment vehicle is stopped in approximately 4.5 metres of pellets of expanded polystyrene and experiences a peak deceleration rate of 65g.

Also at NASA Glenn is the 2.2 Second Drop Tower, which has a drop distance of 24.1 metres. Experiments are dropped in a drag shield, in order to reduce the effects of air drag. The entire package is stopped in a 3.3 metre tall air bag, at a peak deceleration rate of approximately 20g. While the Zero Gravity Facility conducts one or two drops per day, the 2.2 Second Drop Tower can conduct up to twelve drops per day.

NASA's Marshall Space Flight Centre hosts another drop tube facility that is 105 metres tall and provides a 4.6 second free fall under near-vacuum conditions.

Humans cannot utilize these gravity shafts, as the deceleration experienced by the drop chamber would likely kill or seriously injure anyone using them; 20g is about the highest deceleration that a fit and healthy human can withstand momentarily without sustaining injury.

Other drop facilities worldwide include:

- Micro-Gravity Laboratory of Japan (MGLAB) – 4.5 s free fall
- Experimental drop tube of the metallurgy department of Grenoble – 3.1 s free fall
- Fallturm Bremen University of Bremen in Bremen – 4.74 s free fall
- Queensland University of Technology Drop Tower - 2.0 s free fall

Neutral Buoyancy

Weightlessness can also be simulated by creating the condition of neutral buoyancy, in which human subjects and equipment are placed in a water environment and weighted or buoyed until they hover in place. NASA uses neutral buoyancy to prepare for extra-vehicular activity (EVA) at its Neutral Buoyancy Laboratory. Neutral buoyancy is also used for EVA research at the University of Maryland's Space Systems Laboratory, which operates the only neutral buoyancy tank at a college or university.

Neutral buoyancy is not identical to weightlessness. Gravity still acts on all objects in a neutral buoyancy tank; thus, astronauts in neutral buoyancy training still feel their full body weight within their spacesuits, although the weight is well-distributed, similar to force on a human body in a water bed, or when simply floating in water. The suit and astronaut together are under no net force, as for any object that is floating, or supported in water, such as a scuba diver at neutral buoyancy. Water also produces drag, which is not present in vacuum.

Weightlessness in a Spacecraft

Long periods of weightlessness occur on spacecraft outside a planet's atmosphere, provided no propulsion is applied and the vehicle is not rotating. Weightlessness does not occur when a spacecraft is firing its engines or when re-entering the atmosphere, even if the resultant acceleration is constant. The thrust provided by the engines acts at the surface of the rocket nozzle rather than acting uniformly on the spacecraft, and is transmitted through the structure of the spacecraft via compressive and tensile forces to the objects or people inside.

Weightlessness in an orbiting spacecraft is physically identical to free-fall, with the difference that gravitational acceleration causes a net change in the *direction*, rather than the *magnitude*, of the spacecraft's velocity. This is because the acceleration vector is perpendicular to the velocity vector.

In typical free-fall, the acceleration of gravity acts along the direction of an object's velocity, linearly increasing its speed as it falls toward the Earth, or slowing it down if it is moving away from the Earth. In the case of an orbiting spacecraft, which has a velocity vector largely *perpendicular* to the force of gravity, gravitational acceleration does not produce a net change in the object's speed, but instead acts centripetally, to constantly "turn" the spacecraft's velocity as it moves around the Earth.

Because the acceleration vector turns along with the velocity vector, they remain perpendicular to each other. Without this change in the direction of its velocity vector, the spacecraft would move in a straight line, leaving the Earth altogether.

Weightlessness at the Centre of a Planet

The net gravitational force due to a spherically symmetrical planet is zero at the centre. This is clear because of symmetry, and also from Newton's shell theorem which states that the net gravitational force due to a spherically symmetric shell, e.g., a hollow ball, is zero anywhere inside the hollow space. Thus the material at the centre is weightless.

Human Health Effects

Following the advent of space stations that can be inhabited for long periods, exposure to weightlessness has been demonstrated to have some deleterious effects on human health. Humans are well-adapted to the physical conditions at the surface of the Earth. In response to an extended period of weightlessness, various physiological systems begin to change and atrophy. Though these changes are usually temporary, long term health issues can result.

The most common problem experienced by humans in the initial hours of weightlessness is known as space adaptation syndrome or SAS, commonly referred to as space sickness. Symptoms of SAS include nausea and vomiting, vertigo, headaches, lethargy, and overall malaise. The first case of SAS was reported by cosmonaut Gherman Titov in 1961. Since then, roughly 45% of all people who have flown in space have suffered from this condition.

The duration of space sickness varies, but in no case has it lasted for more than 72 hours, after which the body adjusts to the new environment. NASA jokingly measures SAS using the "Garn scale", named for United States Senator Jake Garn, whose SAS during STS-51-D was the worst on record. Accordingly, one "Garn" is equivalent to the most severe possible case of SAS.

The most significant adverse effects of long-term weightlessness are muscle atrophy and deterioration of the skeleton, or spaceflight osteopenia. These effects can be minimized through a regimen of exercise. Astronauts subject to long periods of weightlessness wear pants with elastic bands attached between waistband and cuffs to compress the leg bones and reduce osteopenia.

Other significant effects include fluid redistribution (causing the "moon-face" appearance typical of pictures of astronauts in weightlessness), a slowing of the cardiovascular system, decreased production of red blood cells, balance disorders, and a weakening of the immune system. Lesser symptoms include loss of body mass, nasal congestion, sleep disturbance, excess flatulence, and puffiness of the face. These effects begin to reverse quickly upon return to the Earth.

In addition, after long space flight missions, astronauts may experience severe eyesight problems. Such eyesight problems may be a major concern for future deep space flight missions, including a manned mission to the planet Mars.

On December 31, 2012, a NASA-supported study reported that manned spaceflight may harm the brains of astronauts and accelerate the onset of Alzheimer's disease.

Reduced Muscle Mass, Strength and Performance in Space

There is a growing research database which suggests that skeletal muscles, particularly postural muscles of the lower limb, undergo atrophy and structural and metabolic alterations during space flight. However, the relationships between in-flight exercise, muscle changes and performance are not well understood. Efforts should be made to try to understand the current status of in-flight and post-flight exercise performance capacity and what the goals/target areas for protection are with the current in flight exercise program.

From the very beginning of the U.S. human space program, serious and reasonable concern has been expressed about exposure of humans to the microgravity of space due to the potential systemic effects on terrestrially-evolved life forms that are so suitably adapted to Earth gravity. Humans in the microgravity environment of space, within our current space vehicles, are exposed to various mission-specific periods of skeletal muscle unloading (unweighting). Unloading of skeletal muscle, both on Earth and during spaceflight, results in remodelling of muscle (atrophic response) as an adaptation to the reduced loads placed upon it.

As a result, decrements occur in skeletal muscle strength, fatigue resistance, motor performance, and connective tissue integrity. In addition, there are cardiopulmonary and vascular changes, including a significant decrease in red blood cell mass, that have an impact on skeletal muscle function. This normal adaptive response to the microgravity environment is, for the most part, of little consequence within the space vehicle *per se*, but may become a liability resulting in increased risk of an inability or decreased efficiency in crewmember performance of physically demanding tasks during extravehicular activity (EVA) or abrupt transitions to environments of increased gravity (return to Earth, landing on the surface of another planetary body).

In the U.S. human space program, the only in-flight countermeasure to skeletal muscle functional deficits that has been utilized thus far is physical exercise. In-flight exercise hardware and protocols have varied from mission to mission, somewhat dependent on mission duration and the volume of the spacecraft available for performing countermeasures. Collective knowledge gained from these mission has aided in the evolution of exercise hardware and protocols in attempts to refine the approach to prevention of spaceflight-induced muscle atrophy and the concomitant deficits in skeletal muscle function.

Long duration missions and exploration missions with several transitions between gravitational environments present the greatest challenges to risk mitigation and to development of countermeasures of proven efficacy.

Russian scientists have utilized a variety of exercise hardware and in-flight exercise protocols during long-duration spaceflight (up to and beyond one year) aboard the Mir space station. On the International Space Station (ISS), a combination of resistive and aerobic exercise has been used. Outcomes have been acceptable according to current expectations for crewmember performance on return to Earth. However, for missions to the Moon, establishment of a lunar base, and interplanetary travel to Mars, the functional requirements for human performance during each specific phase of these missions have not been sufficiently defined to determine whether currently developed countermeasures are adequate to meet physical performance requirements.

Access to human crewmembers during both short- and long-duration mission for the study of skeletal muscle adaptation to microgravity and the efficacy of countermeasures has been, and continues to be, limited. Consequently, a more complete understanding of physiologic models for conduct of both fundamental and applied skeletal muscle research. Various models for which sufficient data have been collected have been concisely reviewed. Such models include horizontal or head-down bed rest, dry immersion bed rest, limb immobilization, and unilateral lower-limb suspension. While none of these ground-based analogs provides a perfect simulation of human microgravity exposure during spaceflight, each is useful for study of particular aspects of muscle unloading as well as for investigation of sensorimotor alterations.

Due to limitations in the number of spaceflights and crewmembers in which novel countermeasures can be tested, future development, evaluation and validation of new countermeasures to the effects of skeletal muscle unloading will likely employ variations of these same basic ground-based models. Prospective countermeasures may include pharmacologic and/or dietary interventions, innovative exercise hardware providing improved loading modalities, locomotor training devices, passive exercise devices, and artificial gravity either as an integral component of the spacecraft or as a discrete device contained within it. With respect to the latter, the hemodynamic and metabolic responses to increased loading provided by a human-powered centrifuge have been described recently. Even more recently, an approach to provide both aerobic and resistive exercise by incorporating a cage-like platform into the design has been developed by the same investigator group.

Animal studies, conducted both during spaceflight and in ground-based simulations of the skeletal muscle unloading associated with spaceflight, have contributed to the scientific knowledge base in a manner not totally achievable by means of human spaceflight and ground-based analog studies alone. This is because many of the variables present with human subject investigations can be more tightly controlled in animal studies, and the much larger number of animals typical of such experiments contributes to a greater statistical power to detect differences. A major advantage in use of rodent models is that the adaptive changes to both spaceflight and hind-limb suspension occur in a much shorter time frame than they do in humans (hours to days versus days to weeks). This enables prediction of long-term changes in human skeletal muscle based on the shorter absolute time frame of the rodent investigations. Additionally, it is possible to perform a highly controlled, straightforward experiment in rodents without a requirement to provide some type of countermeasure intervention that introduces a confounding variable. In human studies, it is not possible on ethical grounds to withhold countermeasures known to have some degree of effectiveness to provide a population of true control subjects, in which only the effects of spaceflight are seen, for comparison to subjects utilizing countermeasures modalities. Animal studies do not suffer from such restrictions. Further work is needed to provide a better understanding of the problem, which will allow novel approaches to countering loss of skeletal muscle function associated with spaceflight in humans. Relevant animal spaceflight studies, as well as investigations using muscle unloading paradigms that contribute to our current knowledge base, are presented.

Effects on Non-Human Organisms

Russian scientists have observed differences between cockroaches conceived in space and their terrestrial counterparts. The space-conceived cockroaches grew more quickly, and also grew up to be faster and tougher.

Chicken eggs that are put in microgravity two days after fertilization appear not to develop properly, whereas eggs put in microgravity more than a week after fertilization develop normally.

A 2006 Space Shuttle experiment found that *Salmonella typhimurium*, a bacterium that can cause food poisoning, became more virulent when cultivated in space.

Technical Adaptation in Zero-Gravity

Weightlessness can cause serious problems on technical instruments, especially those consisting of many mobile parts. Physical processes that depend on the weight of a body (like convection, cooking water or burning candles) act differently without a certain amount of gravity. Cohesion and advection play a bigger role in space. Everyday work like washing or going to the bathroom are not possible without adaptation. To use toilets in space, like the one on the International Space Station, astronauts have to fasten themselves to the seat. A fan creates suction that carries the waste away. Drinking is aided with a straw or from tubes.

Bibliography

Adler, R; Bazin M; Schiffer M: *Introduction to General Relativity*, McGraw-Hill Book Company, New York, 1965.

Alexander R. McBirney: *Igneous petrology*, Jones and Bartlett Publishers, Boston, 2007.

Bhagwat, S.B.: *Foundation of Geology,* Global Vision Publishing House, Daryaganj, New Delhi, 2009.

Braja M. Das: *Principles of geotechnical engineering*, THOMSON LEARNING (KY), England, 2006.

Bryn Hubbard, Neil Glasser: *Field techniques in glaciology and glacial geomorphology*, J. Wiley, Chichester, England, 2005.

Charles Lyell: *Principles of geology*, University of Chicago Press, Chicago, 1991.

Dalrymple, G.B.: *The Age of the Earth*, Stanford University Press, California, 1991.

David M. Kipping: *The Transits of Extrasolar Planets with Moons*, Springer, 2011

Eckey, Bernard: *Advanced Soaring Made Easy*,Eqip Verbung & Verlag GmbH, 2007.

Einstein, A: *The Meaning of Relativity,* Princeton University Press, Princeton,, New Jersey, 1956.

Feynman, R. P.; Morinigo, F. B.; Wagner, W. G.; Hatfield, B.: *Feynman lectures on gravitation*, Addison-Wesley, 1995.

Frank S. Spear: *Metamorphic phase equilibria and pressure-temperature-time paths*, Mineralogical Soc. of America, Washington, DC, 1995.

Grimshaw, R.: *Environmental Stratified Flows*, Kluwer Academic Publishers, Boston, 2002.

H. Robert Burger, Anne F. Sheehan, Craig H. Jones: *Introduction to applied geophysics : exploring the shallow subsurface*, W.W. Norton, New York, 2006.

Holmes, Arthur: *Principles of victoria Physical Geology,* Thomas Nelson & Sons, Edinburgh, 1944..

Jacobson, M.: *Fundamentals of Atmospheric Modeling*, Cambridge University Press, Cambridge, UK, 1999.

Lanczos, C.: *The Variational Principles of Mechanics,* Dover Publications, New York, 1986.

Le Grand, Homer Eugene: *Drifting Continents and Shifting Theories*, Cambridge University, 1988.

McElhinny, Michael W.; McFadden, Phillip L.: *Paleomagnetism: Continents and Oceans*, Academic Press, 2000.

Merrill, Ronald T.: *Our Magnetic Earth: The Science of Geomagnetism*, University of Chicago Press, 2010.

Nappo, C.: *An Introduction to Atmospheric Gravity Waves*, Academic Press, Boston, 2002.

Oreskes, Naomi: *The Rejection of Continental Drift*, Oxford University Press, London, 1999.

Pais, A.: *Subtle is the Lord: The Science and the Life of Albert Einstein*, Oxford University Press, 1982.

Pauli, Wolfgang Ernst: *General Theory of Relativity*, Courier Dover Publications, New York, 1958.

Rastall, Peter: *Postprincipia: Gravitation for Physicists and Astronomers*, World Scientific, New York, 1991.

Richard C. Selley: *Elements of petroleum geology*, Academic Press, San Diego, New York, 1998.

Swerdlow, Noel M.; Neugebauer, Otto: *Mathematical astronomy in Copernicus's De revolutionibus*, Springer-Verlag, 1984.

Thornton, Stephen T.; Marion, Jerry B.: *Classical Dynamics of Particles and Systems,* Brooks Cole, 2003.

Turner, B.: *Buoyancy Effects in Fluids*, Cambridge, Cambridge University Press, UK, 1979.

Weinberg, S.: *Gravitation and Cosmology*, John Wiley and Sons, New York, 1972.

Whittaker, ET.: *A Treatise on the Analytical Dynamics of Particles and Rigid Bodies, with an Introduction to the Problem of Three Bodies,* Dover Publications, New York, 1937.

William D. Nesse: *Introduction to optical mineralogy*, Oxford University Press, New York, 1991.

Zee, A.: *Quantum Field Theory in a Nutshell*, Princeton University Press, Princeton,, New Jersey, 2003.

Index

❑❑❑